TRANSACTIONS

OF THE

AMERICAN PHILOSOPHICAL SOCIETY

HELD AT PHILADELPHIA
FOR PROMOTING USEFUL KNOWLEDGE

NEW SERIES—VOLUME 58, PART 6
1968

THE SYSTEMATICS AND EVOLUTION
OF THE MOINIDAE

CLYDE E. GOULDEN

Associate Curator of Limnology, Academy of Natural Sciences of Philadelphia

---

THE AMERICAN PHILOSOPHICAL SOCIETY
INDEPENDENCE SQUARE
PHILADELPHIA

October 1968

To
TED F. ANDREWS
who introduced me to
the Cladocera

Library of Congress Catalog
Card Number 68–54558

PREFACE

Moina belongs to that very interesting group of organisms found in temporary pools and saline lakes. The species are adapted to survive frequent dry periods and to populate rapidly newly formed pools. *Moina* shares this unusual habitat with other branchiopod Crustacea, the Conchostraca, Notostraca, and the Anostraca.

Moreover, the genus is also of significance because the north-temperate species have frequently been used as experimental animals in physiological studies as well as in studies by embryologists and geneticists—most notably in the extensive works of Weismann (1877*a* and *b*, 1879*a* and *b*, etc.) and of Banta (1939) and his associates. Furthermore, there is every likelihood that species of the genus will continue to be used as experimental animals; for they are easily maintained in the laboratory and display interesting adaptive phenomena such as rapid development, ability to co-exist with other species of the genus, and an interesting sexual isolating mechanism.

A critical review of the genus has long been overdue. Guerne and Richard (1892) in the late nineteenth century first spoke of the need for a revision; many investigators have voiced the same opinion, but few have done more. Brehm (1933) attempted a review of the group, but in his study he merely listed the species and gave a few diagnostic characters for some of them.

Many of the features used for definition of species are subjective, such as form of the body or head. Some species have been described on the basis of one or two characters that are peculiar to individuals and not to populations while others were differentiated from pre-existing species because the investigator discovered a feature which though widespread in the genus did not happen to be known to him. Failure to compare specimens from various parts of the world, resulted in the description of new species which, in fact, were based only on regional or subspecific differences.

Perhaps the most noteworthy study of *Moina* is by Gauthier (1954), limited, however, to three species of the genus. Even here, Gauthier failed to recognize certain important species characters.

In the present study I have attempted to use all possible characters that might help to define the species. Whenever possible, I have used type material for the re-description of species; and in all instances, I have referred to the original description. If type material was unavailable, topotypic specimens were sought. However, for most species described prior to 1900, no type material is available nor is it possible to obtain topotypic material. This is particularly true of the forms described by Jurine in 1820. In this instance satisfactory identifications have been possible only by a process of elimination.

Material for all species, in addition to the original types, and topotypes, will be found either in the United States National Museum, the British Museum (Natural History), or the Zoologisk Museum of the University of Oslo. Location of type material is stated in the text description of each species.

After delimitation of the species, it was evident that all forms fell naturally into two distinct groups, presumably representing phylogenetic lines. To recognize the relationships between these two lines, it was necessary to compare characteristics of *Moina* with other cladoceran genera, particularly those of the Daphniidae to which family *Moina* has been referred, as well as with certain genera of the Macrothricidae and Sididae.

In this comparative study, I found that not only is *Moina* related to the daphniid genera but is also related to the sidids, as has already been suggested (Cannon, 1933), as well as to the macrothricids. *Moina*, in form and structure has indeed many characteristics in common with the macrothricids and sidids—characters that presumably were shared with the primitive cladoceran ancestor. However, *Moina* is a specialized form itself and very different from the other groups. For such reasons, given later in detail, I propose to place *Moina*, and the closely related *Moinodaphnia*, in a distinct family, the Family Moinidae.

3

ACKNOWLEDGMENTS

It is very difficult to list all of the people who have contributed to this work without missing some. As in any large work, this study could not have been completed without the help of many friends and interested persons who contributed both time, ideas, and collections of *Moina*. First, however, I want to mention the help of my wife, Nancy Ann Goulden, who has read many drafts of individual chapters.

Professor G. Evelyn Hutchinson has not only contributed encouragement but has also taken time to read and criticize the manuscript. To him I owe thanks for financial support and for the use of equipment. Furthermore, Professor Hutchinson made available plankton collections from many parts of the world.

The Biology Department of Yale University, and particularly its former chairman Professor Donald F. Poulson, have been generous with working space and equipment. The very complete library of Yale University considerably reduced the time required for library research.

Three individuals have helped immensely in criticizing the manuscript and in supplying animals from their personal collections. These are Dr. John L. Brooks, Dr. Ursula M. Cowgill, and Dr. David G. Frey.

Dr. Brooks has willingly placed at my disposal his library of reprints and plankton collections. At the time this study was undertaken, Dr. Brooks had in his possession the large collections of S. F. Light and of E. A. Birge which Dr. Brooks kindly allowed me to use.

Dr. Olga Sebestyen kindly translated Daday's (1888) Hungarian description of *Moina salina* and aided me greatly during my visit to Hungary. Dr. Livia Pirocchi Tonolli sent pertinent references about *Moina* in Italy, and Drs. Ian Thomas, Ian Bayly, A. Moroni, and F. D. Mordukhai-Boltovskoi, at some expense, sent specimens to me from different parts of the world for which I am very grateful.

Many other people have contributed plankton collections of *Moina*. This list is too long to include here but I have attempted to acknowledge them in the text.

This study would have been incomplete without the aid of personnel of various museums who helped to find type material of the described species of *Moina*. The following curators of collections were of immeasurable help in this regard: Dr. H. K. Farkas, Magyar Nemzeti Muzeum, Termeszettudomany Museum, Budapest, Hungary; Dr. H. Gruner, Zoologische Museum, Humboldt University, Berlin (East); Dr. J. P. Harding, Keeper of Zoology, British Museum (Natural History), London, England; Dr. G. Hartmann, Zoologisches Museum, University of Hamburg, Germany; Dr. Nils Knaben, Zoologisk Museum, University of Oslo, Norway.

I have received financial support for this study from several sources. The work was completed at Yale University and was largely supported by National Science Foundation Grants Numbers 1781 and B473, both awarded to Professor G. Evelyn Hutchinson.

Travel expenses for visits to European museums were most kindly supplied by the American Philosophical Society and the Society of The Sigma Xi.

I should like to thank Dr. Raul S. Olivier for permission to duplicate his illustrations and text descriptions of *Moina eugeniae*, and the Akademische Verlagsgesellschaft of Leipzig for permission to reproduce Brehm's illustrations and figures of *Moina hutchinsoni* from Vol. 117 of Zoologischer Anzeiger. Mr. Stanley F. Love, Sales Manager of the College Department, Rand McNally and Co. kindly gave permission to duplicate the text maps of Goode's Interrupted Homolosine Projection of the World.

I am very grateful to Dr. Allison R. Palmer of the United States Geological Survey for permission to describe and illustrate the fossil ephippium of *Moina* recovered by him from the Miocene deposits of the Mojave Desert, California.

C. E. G.

THE SYSTEMATICS AND EVOLUTION OF THE MOINIDAE

CLYDE E. GOULDEN

CONTENTS

I. INTRODUCTION

The first species now referred to the genus *Moina* were described by Straus (1819, 1820) and by Jurine (1820) and put by them in the genera *Daphnia* and *Monoculus* respectively; *Monoculus* has subsequently been removed from availability (Fox, 1951; Hemming, 1958: Opinion 288). The genus *Moina* was later described by Baird (1850) and classed in the Family Daphniidae. Baird also included in this family the genera now belonging to the families Sididae and Macrothricidae later established by Sars (1865; the Family Macrothricidae was originally named Lyncodaphnidae by Sars).

Moinodaphnia was not described until 1887 by Herrick and because it possessed an ocellus and a supposed abdominal process, it was considered to be a possible "missing link" (Herrick, 1887: pp. 35–36) between *Moina*, which lacked these characters, and the other daphniids (*Ceriodaphnia*, *Daphnia*, *Scapholeberis*, and *Simocephalus*), all of which had an ocellus and an abdominal process.

It has therefore been reasoned that, since *Moina* and *Moinodaphnia* possess the daphniid type of thoracic limbs and since *Moinodaphnia* has the typical daphniid features absent in *Moina*, the moinids must be closely related to the other genera in the Daphniidae. However, this reasoning is not completely sound for *Moinodaphnia* proves to be a very specialized form of *Moina* and, although it has an ocellus, the supposed abdominal process is a broad fold in the abdominal exoskelton, not strictly homologous with the structure in other daphniid genera.

Consequently, the evidence for retaining *Moina* and *Moinodaphnia* in the Family Daphniidae rests solely on the similarity in structure of the thoracic limbs. Comparison of the external morphology of other anomopoda and ctenopoda Cladocera reveal many common characteristics between the sidid, macrothricid, and moinid Cladocera. Moreover, comparison of the moinid trunk limbs with those of the daphniids reveals several differences in the first and fifth pair of limbs. Variations in structure of the trunk limbs of the moinids and daphniids suggest that the former group has been separated from the main daphniid line for a considerable period of time, while the external features of the moinids indicate that they have retained many features shared with other families of the Cladocera and absent in the daphniids. Presumably, these characters were also shared by a common ancestor of the Cladocera.

In addition, all species of *Moina* and *Moinodaphnia* have a "Nährboden" or placenta. This is a unique feature of the group—only the polyphemid Cladocera (if these actually are Cladocera) also possess an analogous structure.

These differences are important and indicate the necessity for a reappraisal of the moinids, which obviously must be considered as either a distinct subfamily of the Daphniidae or given family rank themselves. An analysis of characteristics and relationships between families of the Cladocera suggests that the latter procedure is the more satisfactory. Justification for this conclusion is given below.

DIAGNOSIS OF THE FAMILY MOINIDAE

The following features serve to define the Family Moinidae:

1. Five trunk limbs. Limb one of the female completely lacks an exopod; though this structure is present on the first leg of males of some species. Limbs one and two are specialized for cleaning the shell and the other limbs; limbs three and four are filter appendages; limb five is greatly reduced and only aids in producing a water current. A maxillary process is present only on limb two.

2. An ocellus is absent in all forms except *Moinodaphnia macleayi* and one species of *Moina*.

3. The antennules of the female arise from the flat ventral surface of the head, never behind a ventral protuberance as in the Daphniidae, nor on the anterior end of the head as in the Macrothricidae. The male antennules, however, do originate from the anterior end of the head. The antennules are usually ringed with rows of short setae and have a vertical row of long hairs on the lateral margin.

4. All species have a set of muscles attached at the inner margin of the exoskeleton above the eye. In several instances, these muscles may depress the exoskeleton to form a supra-ocular depression.

5. Most species have hairs on the surface of the head, or on the shell, or both. These hairs may be restricted to a small area of the head or shell.

6. The abdominal process of the daphniids is completely absent in the moinids. Some forms have a horseshoe-shaped abdominal fold that holds the embryos in the brood sac, but this fold is distinct from the abdominal process found in the Daphniidae.

7. The adult parthenogenetic females have a placenta or "Nährboden."

8. The postabdomen has only a row of lateral teeth; there are no marginal teeth. The distal-most tooth of the lateral row is usually bident while the remaining teeth are feathered with fine setae.

9. The postabdomen claw is smooth or has a pecten but never has only one or two basal spines.

10. A pair of hooks is present at the dorsal end of the posterior shell margin which hold the abdominal setae of the postabdomen.

11. The ephippium has one or two sexual eggs and is usually reticulated over its entire surface.

12. The antennules of the male have two sensory setae, one is short, with a broad base and is situated on the medial margin at a point where the antennules are bent. The second seta is long, thin, and usually on the lateral margin.

13. The male antennules are long and broadly curved and are thus modified for clasping the female.

14. The genital opening is always located on the ventral side of the male postabdomen, never on the dorsal side.

Genera:

Moina Baird, 1850
Moinodaphnia Herrick, 1887

PHYLOGENETIC RELATIONSHIPS AND CHARACTERISTICS OF THE MOINIDAE

The Moinidae are easily identified by their rather prominent head with the pair of long, thin "cigarette-shaped" antennules that extend from the ventral margin of the head (fig. 1, *A*). The postabdomen further characterizes the animal by its large post-anal extension that bears a row of lateral feathered setae and one distal bident tooth. The Moinidae are also well characterized by their habitat; the majority of species are restricted to small temporary pools or saline and alkaline lakes. They are seldom found in permanent ponds and lakes.

MORPHOLOGY OF THE FEMALE HEAD

The form of the moinid head is basically similar to that of the Sididae and the antennules originate from the ventral surface of the head, just below the eye. In contrast, the antennules of the daphniids are short and always originate near the labrum, behind a protuberance of the head, while the antennules of the macrothricids, which are usually as long as those of *Moina*, are located at the anterior end of the ventral head margin.

The moinids have a large fused compound eye but normally lack the ocellus in the genus *Moina*, although it is present in one species of *Moina* (*reticulata*) and in *Moinodaphnia macleayi*. This is the only major group of the Cladocera that commonly lacks an ocellus (the Leptodoridae and Polyphemidae also lack an ocellus, but I am of the opinion that these families may not belong to the Cladocera).

The head is frequently indented above the eye (the so-called supra-ocular depression) by the attachment of a muscle bundle to the inner surface of the flexible exoskeleton. The muscle bundle appears to consist of a pair of muscles from the intestine and the levator muscle from the labrum (see Binder, 1932, for comparable condition in *Daphnia*).

The second antenna consists of a large basipod, an endopod of three segments and an exopod of four segments (fig. 1, *B*). The basipod has two sensory setae that originate from the posterior margin near the base of the segment. There is a third sensory seta on the distal end of this segment between the two rami. The precise function of these setae is unknown, but it has been suggested that they may serve to detect the movement of water past the body (Hantschmann, 1961). Although the setae are present on the second antennae of all Cladocera they are seldom so well developed as in the Moinidae. *Moina* would be an excellent experimental animal for study of the function of these setae.

Each segment of the second antennal endopod bears one or more long swimming setae on its distal end. The first two segments each have one while the distal segment has three long setae and a single short spine.

The four-segmented exopod has long swimming setae only on the two distal segments, while the short basal segment lacks setae and the second segment has only a short spine. The third segment has one long seta and the distal segment has three long swimming setae and a short spine. In *Moinodaphnia*, the spine on the terminal segment, homologous to the short spine of *Moina*, is very long, as long as the distal swimming hairs so that there appear to be four distal swimming setae.

The exopod also bears an inner row of short teeth which probably aid in cleaning detritus from the surface of the shell.

The segmentation and setae pattern of the moinid second antenna is comparable to that found in the daphniids and in many macrothricids. The second antennae of the sidid Cladocera differ considerably from the other groups because they have only three segments in the distal rami but each segment usually bears several swimming setae.

THORACIC LIMBS

The trunk limbs of the Cladocera are basically filtering appendages as is true of most branchiopods. Because the limbs are the organism's feeding structures, alterations in limb structure must be very conservative. Differences that develop should contribute to an increase in efficiency of the feeding process. An understanding of this concept is necessary for consideration of affinities within the Cladocera and particularly so, in the present context, for recognition of the proper phylogenetic position of the moinids.

There is general agreement that the sidid limbs represent the primitive condition in the Cladocera (Cannon, 1933) although Eriksson (1934) has suggested that the sidids and daphniids had separate origins from the Conchostraca. The sidids have retained six pairs of almost identical limbs; each limb has a basipod, an endopod, and an exopod. The basipod has an epipodite, sometimes referred to as the branchial sac, and a gnathobase (Lilljeborg, 1900). The endopod is a tubular segmented branch with a large filter comb; the exopod is a large flabelliform structure. The gnathobase of each limb pushes food forward in the food groove; the comb on the endopod filters suspended particles from the water, and the exopod closes off the lateral side of the filter chamber (Cannon, 1933). The sixth pair of limbs is reduced in size and may, besides its filtering function, close off the posterior end of the filter chamber.

In the moinids and the daphniids the number of true filtering appendages has been reduced to two (Cannon, 1933). Only five limbs are present, and each limb has been specifically modified towards greater efficiency in its particular function.

The first pair of limbs are no longer filtering limbs. They are considerably reduced in size and have far fewer setae than found on the sidid first limb. The basipod segment lacks a gnathobase but has the epipodite (fig. 1, *C*). The second and third segments have endites with posteriorly directed setae. There are two endites on the second segment, the first endite has three setae in *Moina* (four in *Daphnia*) and the second endite has two setae (three in *Daphnia*). The third segment has one posterior and one anterior seta in *Moina* while in *Daphnia* there are two posterior setae and one short lateral seta but no anterior seta. Finally, the fourth and distal segment of *Moina* has three terminal setae while *Daphnia* has only one. The first limb of *Daphnia* retains an additional distal segment that appears to be homologous with the exopod of the male first leg (fig. 1, *D* and *E*). This exopod is absent on the legs of female moinids but remains on the first leg of the males of a few species (fig. 1, *F*).

Although the first limb of *Moina* has lost its filtering function, it is still present in the adult organism and therefore it must be important to the animal. Its present function is not known but the distal setae of this limb probably clean detritus from the inner shell surface or from the other appendages. Furthermore, it is probably a necessary aid in production of the feeding current, for it is apparent in live animals that this limb moves synchronously with the other limbs. Cannon (1933) has suggested that its movements produce a forwardly directed current in the food groove which would move the food to the mouth. However, such a current has not been found in other Cladocera (see Fryer, 1963: p. 376, for a discussion of this current).

The second limb, as the first, has lost the filter comb and the filtering function. The basipod of this limb has retained a lobe with a long row of setae which seems to be homologous to the gnathobase of the Sididae but functions only in part as a true gnathobase. In both *Moina* and *Daphnia* this "gnathobase" is greatly enlarged and has three or four proximal setae that are recurved and which may move food forward in the food groove. The other setae are modified for scraping food particles from the third limb (Cannon, 1933). The two distal setae in the "gnathobase" reach to the back of the food groove and are modified for scraping food particles from the posterior limbs, and may also help move food from the posterior end of the groove. This function has not been demonstrated; but if Cannon's proposed forward-directed current in the food groove is not actually present, then some other mechanism must be in effect. These setae extend down into the food groove and might well, during normal movement of the second limb,

FIG. 1. General features of the Moinidae. *A. Moina brachiata* female, Slaghille, Denmark, 4-IX-26 (collected by K. Berg). *B. Moina micrura*, ventral view of second antenna of female from Mare T. Vienp. et Briqueterie, Port au Prince, Haiti, 1-VI-1896 (Birge Collection No. 800 from Richard's collections). *C. Moina wierzejskii*, female first leg, roadside ditch along county road one mile east of Cheyenne Bottoms Waterfowl Refuge, Barton County, Kansas, 26-VII-58. *D. Daphnia pulex*, medial view of female first leg, Clear Lake, California, July, 1964. *E.* Lateral view of *D. F. Moina macrocopa*, male first leg, roadside ditch along county road one mile east of Cheyenne Bottoms Waterfowl Refuge, Barton County, Kansas, 26-VII-58. *G. Moina belli*, female fifth leg, water tank, Aden, 7-XII-32. *H. Moina macrocopa* male, roadside ditch along county road one mile east of Cheyenne Bottoms Waterfowl Refuge, Barton County, Kansas, 26-VII-58.

move the food. Fryer (1963) was able to determine that the tips of the long filtering setae of *Eurycercus lamellatus* aid in moving food forward. The food is then collected by the scrapers and collecting spines of trunk limb two (Fryer: p. 373). It should be obvious that further work is necessary before this process can be well understood in the daphniid and moinid Cladocera.

The gnathobase of thoracic limb two has, therefore, assumed the function of the gnathobase of limb three and in part of limb four (Cannon, 1933). The gnathobases of the latter two limbs have been lost.

Limbs three and four both have a large filter comb on the endopod while the exopods are large and flabelliform and close off the filter chamber laterally as in the sidids. These two limbs lack a gnathobase (Cannon's view that the filter comb of *Daphnia* represents an enlarged gnathobase seems unlikely because such a structure would merely replace the already existent identical filter comb of the endopod of sidid limbs).

Finally, the fifth limb has been considerably reduced in complexity and, in *Moina*, consists of a one or two segmented endopod (the actual number of segments is difficult to discern) and a large flabelliform exopod (fig. 1, *G*). In *Daphnia*, the exopod of limb five has been further reduced in size. The endopod of the fifth limb, in these two forms, now serves primarily in closing off the posterior end of the filter chamber while the exopod seemingly aids in pushing water forward through the filter comb of leg four (Storch, 1924) or may aid in producing a forwardly directed current in the food groove (Cannon, 1933) if such a current actually exists.

These differences in limb structure represent a very important development in the evolution of the Cladocera. Not only do we have a highly efficient filter feeding organism; but, by reducing the number of limbs necessary for filtering food, the other limbs may either be lost or specialized anew for different functions. The latter situation appears to have resulted in the development of the macrothricid and chydorid Cladocera. These forms, whose limbs are basically similar to the moinid-daphniid limbs, are adapted to live in the benthic mud or in the littoral weeds. In the chydorid, *Eurycercus lamellatus*, the first pair of limbs are modified to clasp vegetation while the second pair of limbs bear stout spines that scrape vegetation from plant stems. The third and fourth limbs then filter the edible food material (Fryer, 1963).

However, the macrothricid and chydorid limbs differ from the moinid and daphniid limbs in having an inner lobe (Fryer, 1963) or maxillary process (Lilljeborg, 1900) that seems to be a rudiment of the sidid gnathobase (this has not been definitely confirmed, however). If this is indeed the gnathobase,

then it would indicate that the macrothricids and related forms did not originate directly from the moinid-daphniid line but rather branched off this line very early in the development from the primitive cladoceran group. Nevertheless, the initial modification of trunk limbs that ultimately led to the moinid and daphniid lines appears to have been necessary for the evolution of the macrothricid and chydorid Cladocera.

POSTABDOMEN

The postabdomen of *Moina* has a large post-anal extension that is conical and ends with the distal claws. There is a row of lateral, feathered setae and a single distal bident tooth on the postabdomen. The claw lacks basal spines but is usually pectinate as in the daphniids.

In the daphniids, the anus is sub-terminal and there may be a rather short post-anal extension. The macrothricids usually have a large post-anal extension.

BROOD–CHAMBER AND RETENTION OF EMBRYOS

The moinids among the anomopod Cladocera are unique in the possession of a placenta or "Nährboden" that supplies nourishment to the developing embryos (Weismann, 1877a). This structure was first described by Weismann, but unfortunately nothing has been done subsequently to learn more of its specific function.

There are no abdominal processes in the moinids. These processes are common in the daphniids (Brooks, 1957) and serve to hold the embryos in the brood pouch (Brooks, personal communication; Banta, 1939: p. 183). Occasionally, as in *Moinodaphnia*, there may be a horseshoe-shaped fold of the abdomen exoskeleton that serves to delimit the posterior end of the brood pouch and may aid in retention of the embryos. There is, however, a device that may serve a similar function as the daphniid abdominal process. At the dorsal end of the posterior shell margin there is a pair of hooks that extend medially. In preserved specimens of *Moina*, the long abdominal setae are usually caught by these hooks. These setae, thus caught, close off the posterior part of the brood pouch and thereby may aid in securing the eggs and embryos in the brood pouch. Comparable hooks may be found on the shells of *Diaphanosoma* (Brehm, 1933). It is possible, therefore, that they first developed in the Sididae and were then retained in the moinids.

THE MALE HEAD

The head of the male moinids is disproportionately large and contains a large compound eye that usually fills the tip of the head. However, the most characteristic feature of the head is the very long pair of antennules that originate from the anterior end of the ventral margin of the head (fig. 1, *H*). These antennules are modified for clasping the female during copulation (Weismann, 1879a). The distal end of each antennule bears three to six large curved hooks that clasp the female's shell. The antennule has two sensory setae that usually originate near the head. One seta is long and thin and originates on the lateral margin of the antennule. It seems to be the counterpart of the female's sensory seta. The second seta is short with a very thick base, and is on the medial margin of the antennule.

This type of male antennule may also be found in the Sididae and Macrothricidae but is completely absent in the Daphniidae. In the daphniids the males have relatively short straight antennules that are usually folded forward. The two sensory setae are located distally. One of these setae is represented by a large peglike seta on the extreme distal end of the antennule while the second seta, which is much smaller than the other, originates slightly proximal. The antennule of Simocephalus differs somewhat from this pattern but is basically the same (Lilljeborg, 1900).

Since the male antennules of the moinids are similar to those of the Sididae and the Macrothricidae, it would suggest that this is a feature shared with a common ancestor of the groups.

FIRST THORACIC LIMB OF THE MALE

The first thoracic limbs of the males of all Cladocera are modified for clasping the shell of the females. These legs carry a large hook on the distal end of the leg that is used to secure the female during copulation. In the Sididae the first leg of the male is only slightly modified by the addition of a fingerlike hook on the distal end of the endopod. The leg maintains its filtering function. In the Anomopoda Cladocera, where the first leg is no longer needed in filtering food, the first leg of the male has been greatly modified to serve its clasping function. Comparison of male and female first legs of Moina (fig. 1, C and F) demonstrates the degree of modification. Although the segmentation and setation pattern of the female first leg may be traced in the male leg, the addition of a large hook as an outgrowth from the third segment dominates the distal end of the leg. The terminal segment with its three setae is greatly reduced in size. In some moinids and in all other Anomopoda Cladocera, there is a long fusiform segment with a very long distal seta that stems from the lateral surface of the limb. This segment represents the remnant of the exopod. The exopod is absent on the legs of the moinid females but is retained in females and males of all other Anomopodan Cladocera.

POSTABDOMEN OF THE MALE

The postabdomen of the moinid males is only slightly modified for its function in sperm transfer.

The armature of the postabdomen is not altered nor is that of the claw. The sperm duct opens near the ventral side of the postabdomen but the precise position of the opening varies. Usually the opening is just proximal and ventral to the row of feathered setae, but in two species it opens at the distal end of the postabdomen ventral to the base of the claw. Kurz (1874) and Gruber and Weismann (1880) described the opening of Moina micrura and rectirostris (=brachiata) as lying near the bend of the postabdomen and abdomen. I have not found the opening in the latter species, but in micrura it is located near the row of feathered setae.

The somewhat proximal opening near the ventral margin of the postabdomen is similar to that found in the sidids. Furthermore, except in those species that have the sperm duct opening near the claw, the sperm duct appears to be merely a saclike extension of the testes on the surface of the postabdomen rather than incorporated within the postabdomen.

In the macrothricids the sperm duct opens on the ventral side of the postabdomen near the claw base. In the daphniids, however, this duct only opens on the dorsal side of the claw near its base. The significance of the different positions of the opening of the sperm duct is unknown, but it must be important in sperm transfer.

SEXUAL CYCLE IN MOINA

The sexual females of Moina, on maturity, produce one or two sexual eggs (one in each ovary) that remain in the ovaries until they are fertilized.[1] The ephippium develops simultaneously with the development of the sexual egg—it is not dependent upon fertilization of the egg by the male. If the female is not fertilized the egg disintegrates and the ephippial shell is cast off (Weismann, 1879a; Dehn, 1948). According to Weismann (1879b) females of Moina continue to produce sexual eggs and ephippia throughout their life. This has not been confirmed by subsequent investigators who instead have found that whereas sexual egg production in a given individual always precedes and never follows the production of parthenogenetic eggs, parthenogenetic broods can be produced in the molt immediately following if the sexual egg is not fertilized (Allen and Banta, 1929; Dehn, 1948; Grosvenor and Smith, 1913).[2]

---

[1] Weismann (1879a) states that the sexual egg remains in the ovary in Moina paradoxa (=macrocopa) until it is fertilized while in rectirostris (=brachiata) the egg is deposited in the brood chamber prior to fertilization. However, I have found that in micrura, a closely related species to brachiata, the egg remains in the ovary as in macrocopa. This is also true of wierzejskii and affinis.

[2] In the four species that I have studied the females must pass through at least one molt following the sexual molt before they produce a parthenogenetic brood. Presumably the placenta develops during the post-sexual instar.

In the daphniid Cladocera, it appears that any parthenogenetic female may produce sexual eggs (Berg, 1931). The *Moina* females are strikingly different in this regard for only newly matured females may produce sexual eggs (Weismann, 1879*b*; Grosvenor and Smith, 1913). I would suggest that this difference in *Moina* is due to the presence of the "Nährboden" in parthenogenetic females. Sexual females lack a "Nährboden," but the sexual eggs, which have a large supply of yolk, do not complete their development in the brood pouch.

CONCLUSIONS

The external morphology of the moinids, particularly of the head and its appendages in the males and females, is strikingly similar to that of the Sididae and the Macrothricidae and, except for the second antennae, quite unlike that found in the Daphniidae. It would seem reasonable to assume that the head morphology is secondarily modified in the Daphniidae while the other groups have retained the condition shared with an ancestral cladoceran group.

On the other hand, the structure of the moinid limbs clearly suggests that they are related to the daphniid Cladocera, although the first and fifth limbs of the moinids are different and suggest a rather long period of separation from the main daphniid line.

Furthermore, the moinids are unique among the Cladocera (except for the Polyphemidae; Weismann, 1877*a*) in possessing a "Nährboden" to nourish the parthenogenetic young and in having an unusual sexual reproduction cycle.

The present evidence suggests that the anomopod Cladocera developed from a group of ctenopodlike Cladocera that possessed rather identical filtering limbs. The tendency towards increased efficiency in filtering food permitted a reduction in number of limbs necessary for the feeding process. This, in turn, allowed accessory limbs to either disappear or to respecialize for other functions. One early branch from this line of development gave rise to the macrothricids and related families (Bosminidae and Chydoridae) while the tendency towards further filtering efficiency eventually gave rise to the moinid and daphniid Cladocera.

The moinids, however, have retained a sufficient number of features similar to the macrothricids and the sidids and have developed several specializations in adapting to their peculiar habitat to suggest that they formed an early branch from the main daphniid line of evolution. It is possible that they have survived to the present only because of their high degree of physiological adaptation to very temporary water bodies. Such a harsh environment would reduce the intensity of competition that such "primitive" forms might otherwise encounter in permanent habitats containing many species and would consequently improve the chances of their survival.

II. CHARACTERISTICS OF THE GENUS MOINA

The genus *Moina* was described by Baird (1850) to include two species, *Monoculus rectirostris* Jurine (1820) and *Monoculus brachiatus* Jurine (1820). His description of the genus was brief and included the following comments:

Head rounded and obtuse. Superior antennae of considerable length, one-jointed, arising from the front of the head, near the centre. Inferior antennae very large, and fleshy at the base. (Baird, 1850: p. 100.)

Because he unwittingly described the same species as both *Moina rectirostris* and *M. brachiata*, the generotype is by monotypy, inevitably this species, must be designated *Moina brachiata* (see below, pp. 16–17).

GENERIC DIAGNOSIS

The long movable antennules of the female originate on the flat ventral surface of the head. The eye is located in the center of the head or near the anteroventral margin. An ocellus is rarely present. Most forms have a supraocular depression that is formed by the attachment of a muscle to the inner surface of the exoskeleton above the eye. The female antennules have a single sensory seta on the anterior margin. The antennules frequently have encircling rows of short setae and a vertical row of long hairs. The second antennae are very pubescent. The distal segment of each ramus has three long swimming setae and one very short spine.

The head, shell, or both, are often covered with hairs.

The postabdomen is armed with a usually bident tooth and three to sixteen lateral feathered teeth. The claws may or may not have a pecten but never have single basal denticles.

The ephippium contains one or two sexual eggs. The ephippium develops from the dorsal half of the shell and is ornamented over the entire surface.

The males have long antennules that are modified for clasping the females. There are two sensory setae set either in the middle of the antennule, near the head, or intermediate to these two positions. The distal end of the antennules has three to six recurved hooks. The eye of the male is large and often fills the anterior end of the head.

The second antennae of the male are like those of the female.

The first leg of the male has a medial hook on the distal end of the third segment. The terminal segment is reduced in size and has three setae; one is bare and may be long and curved, as a hook. The other two setae are feathered. The exopod may be present on this leg but usually is not.

The testes are on the sides of the intestine and postabdomen. The genital opening is along the ventral margin of the postabdomen, but the precise position of the opening varies.

The spermatozoa are usually small spherical cells, but sometimes are large.

SPECIES OF THE GENUS MOINA

At least fifty species of *Moina* have been described, and most of these are still referred to in the literature and regarded as valid. Owing to the great morphological plasticity within the genus, many of the described species are actually synonyms or represent geographical subspecies of previously described species. The following list includes all forms described as belonging to the genus *Moina* or to other genera that were subsequently included in *Moina*.

Moina Baird, 1850

> *M. macrocopa* (Straus, 1820)
> *M. brachiata* (Jurine, 1820)
> *M. rectirostris* (Jurine, 1820)
> *M. macleayi* King, 1853
> *M. lemnae* King, 1853
> *M. micrura* Kurz, 1874
> *M. flagellata* Hudendorff, 1876
> *M. fischeri* Hellich, 1877
> *M. lilljeborgi* Schoedler, 1877
> *M. paradoxa* Weismann, 1877
> *M. bathycola* Vernet, 1878
> *M. banffyi* Daday, 1883
> *M. propinqua* Sars, 1885
> *M. salina* Daday, 1888
> *M. azorica* Moniez, 1888
> *M. weberi* Richard, 1891
> *M. dubia* de Guerne et Richard, 1892
> *M. affinis* Birge, 1893
> *M. wierzejskii* Richard, 1895
> *M. australiensis* Sars, 1896
> *M. tenuicornis* Sars, 1896
> *M. weismanni* Ishikawa, 1896
> *M. flexuosa* Sars, 1897
> *M. hartwigi* Weltner, 1898
> *M. minuta* Hansen, 1899
> *M. mongolica* Daday, 1901
> *M. microphthalma* Sars, 1903
> *M. brevicornis* Sars, 1903
> *M. belli* Gurney, 1904
> *M. ciliata* Daday, 1905
> *M. salinarum* Gurney, 1909
> *M. makrophthalma* Stingelin, 1914
> *M. platensis* Biraben, 1917
> *M. minima* Spandl, 1926
> *M. turkomanica* Keiser, 1931
> *M. latidens* Brehm, 1933
> *M. geei* Brehm, 1933
> *M. tonsurata* Brehm, 1935
> *M. esau* Brehm, 1936
> *M. hutchinsoni* Brehm, 1937

> *M. irrasa* Brehm, 1937
> *M. ruttneri* Brehm, 1938
> *M. chankensis* Ueno, 1939
> *M. juanae* Brehm, 1948
> *M. eugeniae* Olivier, 1954
> *M. latirostris* Brehm, 1958
> *M. ganapatii* Brehm, 1963
> *Mediomoina elliptica* Arora, 1931
> *Moinodaphnia brasiliensis* Stingelin, 1904
> *Moinodaphnia reticulata* Daday, 1905

The present study is the first to reduce the number of species in the genus. It should not be surprising therefore that of the fifty species described, only seventeen are identified as valid. Four of the fifty species have been transferred to different genera subsequent to their description. Twenty-six names are considered to be no more than synonyms or deserve only form or subspecies rank, while three names have been placed in *incertae sedis*. At least two of the uncertain species should be considered *nomina dubia*, but all three are probably synonyms of one or more of the other species.

What perhaps may seem incongruous, considering the number of species previously described, has been the necessity to name one new species in the present paper. This new form was collected from a region that has been little studied. It is likely that other new species will be found in other poorly studied regions.

The species recognized in the present study, and the synonyms of each (indented) are given below:

M. brachiata (type species)
> *M. lilljeborgi*
M. macrocopa
> *M. flagellata*
> *M. fischeri*
> *M. paradoxa*
> *M. banffyi*
> *M. azorica*
> *M. esau*
> *M. ganapatii*
M. micrura
> *M. rectirostris*
> *M. propinqua*
> *M. dubia*
> *M. weberi*
> *M. ciliata*
> *M. makrophthalma*
> *M. latidens*
M. affinis
> *M. irrasa*
M. wierzejskii
> *M. platensis*
M. australiensis
M. tenuicornis
M. weismanni
> *M. brevicornis*

M. flexuosa
M. hartwigi
M. minuta
　Moinodaphnia brasiliensis
　Moina minima
M. mongolica (salina)
　M. microphthalma
　M. salinarum
M. belli
　M. tonsurata
　M. turkomanica
　M. latirostris
M. reticulata
　Moinodaphnia reticulata
M. hutchinsoni
M. eugeniae
M. chankensis[3]
M. brachycephala sp. n.
Incertae sedis
　M. geei
　M. ruttneri
　Mediomoina elliptica

Species belonging to other genera:

　M. macleayi (= *Moinodaphnia macleayi*)
　M. lemnae (= *Pseudomoina lemnae*)
　M. juanae (= *Moinodaphnia macleayi*)
　M. bathycola (= *Ilyocryptus acutifrons*; Richard, 1888)

KEY TO THE GENERA AND SPECIES OF THE MOINIDAE

1. Body laterally flattened and shell with a slight dorsal keel; ocellus present　　　　　　*(follow with ¶2)*
   Shell rotund and without a dorsal keel; ocellus usually absent.
   　　　　　　　　　　　　　　(follow with ¶3)

2. Long cigar-shaped antennules on slight protuberance; four distal setae on terminal segments of second antennae exopod (four-segmented ramus).　Circumtropical.
   　　　　　　Moinodaphnia macleayi (pp. 84–87)

   Antennules short and usually folded back against head; body rotund; terminal segment of second antennae with only three long setae; claw with a pecten.　South America only.
   　　　　　　　　Moina reticulata (pp. 72–74)

3. Postabdomen with a distal bident tooth (Key Plate, 1)　(¶5)
   Postabdomen without a distal bident tooth　　　(¶4)
4. Large species (*ca.* 1.2 to 1.6 mm, long) usually with a distal unident tooth on postabdomen; *no* supraocular depression.

---

[3] I have been able to examine only three specimens of *Moina chankensis*, all parthenogenetic females.　The form does appear to be distinct from all other moinids but its affinities to other species cannot be determined without males and sexual females. The three specimens were given to me by Dr. V. Korinek of Charles University, Prague, Czechoslavakia, for which I am very grateful.　Unfortunately the specimens were not available during the course of my studies on *Moina* but were seen only after the manuscript had been completed.
Zooplankton samples containing *Moina chankensis* may be found in the collection of the Zoological Institute of the Academy of Sciences in Leningrad.

Ephippium with two eggs.　Saline lakes in Western North America.
　　　　　　　　Moina hutchinsoni (pp. 74–78)

Small species (*ca.* 1.0 mm. long) lacking distal unfeathered tooth on postabdomen; with a supraocular depression. Ephippium with one egg.　Saline pools in South America.
　　　　　　　　Moina eugeniae (pp. 78–81)

5. First leg of female with anterior seta on penultimate segment. (Key Plate, 2)　　　　　　　　(¶7)
   First leg of female without anterior seta on penultimate segment (Key Plate, 3)　　　　　　(¶6)

6. Large species (1.1 to 1.8 mm. long).　Restricted to Old World saline pools, and lakes.　Ephippium with one egg and ornamented with polygonal reticulations.
   　　　　　　　　Moina mongolica (pp. 64–69)
   Small species (.5 to .7 mm. long) reported only from rivers and lakes from coastal areas of Central and South America. Sensory setae on base of second antennae very long.
   　　　　　　　　Moina minuta (pp. 62–64)

7. First leg of female with anterior setae on penultimate and ultimate segments feathered or only with fine hairs.　(Key Plate, 2)　　　　　　　　　　(¶8)
   First leg of female with anterior setae of penultimate and ultimate segments toothed and not with fine hairs.　Large species with two-egg ephippium that is ornamented with square or polygonal reticulations.　Head usually without a supraocular depression and broad.　Male with exopod on first leg and with antennules bent near middle.　Found in Northern Hemisphere of Old and New World.
   　　　　　　　　Moina macrocopa (pp. 22–28)

8. Long hairs on head and shell of female.　(Key Plate, 4)
   　　　　　　　　　　　　　　(¶9)
   Long hairs only on ventral surface of head or completely absent on female.　　　　　　　　(¶14)

9. Small species, never over 1.2 mm. long, well-developed supraocular depression, narrow head; single egg in ephippium
   　　　　　　　　　　　　　　(¶10)
   Large species, 1.2 to 1.6 mm. long, two eggs in ephippium.
   　　　　　　　　　　　　　　(¶11)

10. Hairs dense on head and shell; setae on posterior shell rim of equal size and ungrouped; postabdomen with long hairs on dorsal margin, claw with distinct pecten.　Ephippium with large round cells.　North America and Italy.
    　　　　　　　　Moina affinis (pp. 37–42)
    Hairs sparse on head and shell; setae on posterior shell rim in groups and unequally sized; postabdomen without long hairs on dorsal margin; claw pecten indistinct.　Ephippium with large round cells.　Far East (China, Japan, Southeast Asia, India).
    　　　　　　　　Moina weismanni (pp. 53–55)

11. Head without a distinct supraocular depression; setae on posterior shell rim of equal size and ungrouped. (Key Plate, 4)　　　　　　　　　　(¶12)
    Head usually with a distinct supraocular depression, setae on posterior shell rim grouped and of unequal size; claw without a distinct pecten.　Ephippium ornamented with round, knoblike projections.　Australia and New Zealand.
    　　　　　　　　Moina australiensis (pp. 48–51)

12. Claw of postabdomen with a distinct pecten.　　(¶13)
    Claw of postabdomen without a distinct pecten.　Male first leg with an exopod and a long distal seta.　Ephippium ornamented with flat, rectangular or polygonal cells.　Africa and Middle East.
    　　　　　　　　Moina belli (pp. 69–72)

13. Antennules of female very long and thin. Ephippium with rectangular reticulations in bricklike pattern. Male first leg with exopod and distal seta. Australia and South Africa.

Moina tenuicornis (pp. 51–53)

Antennules not thin but robust; claw pecten usually very large. Ephippium ornamented with round, knoblike projections. Male first leg without an exopod. Restricted to the New World.

Moina wierzejskii (pp. 42–48)

14. Hairs present on ventral surface of head behind antennules.

(¶15)

Hairs completely absent

(¶16)

15. Postabdomen of female with long hairs on dorsal margin, claw pecten moderately long; setae on posterior shell rim of equal size and ungrouped. Small species 1.1 to 1.2 mm. long. Restricted to Eastern Africa and Madagascar.

Moina hartwigi (pp. 59–61)

Postabdomen without long hairs; claw pecten very large; posterior shell rim with groups of unequally sized setae. Ephippium with embossed globe in center. Old World: Northern Hemisphere and South Africa.

Moina brachiata (pp. 15–22)

16. Head narrow and with a distinct supraocular depression. Less than 1.2 mm. long. (Key Plate, 5)   (¶17)
Head very large and broad and without a supraocular depression. Eye small. Ephippium with two eggs and ornamented with round knoblike cells. Male antennules with three accessory hooks near distal end of antennules. Southern California.

Moina brachycephala (pp. 81–83)

17. Postabdomen with long hairs on dorsal margin, posterior margin of shell with ungrouped short setae. Male antennules bent near head; hairs on shell of male. Flexuous bend in ventral shell rim. Ephippium with knoblike projections. Western Australia.

Moina flexuosa (pp. 55–58)

Postabdomen rarely with long hairs; posterior shell rim with setae in groups. Male antennules bent one-third distance from head; large hook on first leg but exopod absent; spermatozoa round cells with many radiating axons. Ephippium with polygonal cells that are indistinct in middle of ephippium. Throughout tropical and subtropical regions of the world; often found in large permanent lakes.

Moina micrura (pp. 28–37)

III. DESCRIPTION OF THE SPECIES OF MOINA

MOINA BRACHIATA (JURINE, 1820)

Monoculus brachiatus Jurine, 1820: pp. 131–132; pl. 12, figs. 3–4.

Daphnia brachiata Lieven, 1848: pp. 29–30; pl. 17, figs. 7–9.

Moina rectirostris Baird, 1850: p. 101; pl. 11, figs. 1–2.

Moina brachiata Baird, 1850: p. 102; pl. 9, figs. 1–2.

Daphnia rectirostris Fischer, 1851: pp. 105–108; pl. 3, figs. 6–7.

Daphnia brachiata Lilljeborg, 1853: pp. 37–40; pl. 2, figs. 4–5.

Daphnia brachiata Leydig, 1860: pp. 166–174; pl. 4, fig. 39; pl. 5, figs. 40–43.

Moina brachiata P. E. Müller, 1868: pp. 133–134; pl. 2; fig. 22.

Moina brachiata Hellich, 1877: pp. 53–54; fig. 20.

Moina rectirostris Hellich, 1877: pp. 54–55; fig. 21.

Moina lilljeborgi Schoedler, 1877: p. 5; figs. 9a–c, 10.

Moina rectirostris Gruber and Weismann, 1880; pp. 52–81; pl. 3, figs. 3–4; pl. 4, figs. 5, 7, 10; pl. 5, figs. 17, 19, 21; pl. 6, figs. 22–24.

Moina rectirostris Eylmann, 1887: pp. 74–76; pl. 5, fig. 2.

Moina rectirostris Lilljeborg, 1900: pp. 216–222; pl. 29, figs. 23–30; pl. 30, figs. 1–12.

Moina rectirostris Sars, 1903a: p. 179; pl. 7, fig. 5.

Moina lilljeborgi Keilhack, 1914: pp. 150–151.

Moina lilljeborgi var. *salinarum* Rühe, 1914: pp. 16–19; figs. 4e–f.

Moina brachiata Sars, 1916: pp. 321–322; pl. 35, figs. 3, 3a–b.

Moina rectirostris var. *caucasica* Schicklejew, 1930: pp. 338–339; fig. 2.

Moina rectirostris Wagler, 1937: p. 36; figs. 115, 116a–c.

Moina n. sp. (?) Šrámek-Hušek, 1940: p. 212; figs. 2a–f.

Moina rectirostris Behning, 1941: pp. 156–160; figs. 62a–d, 63.

Moina rectirostris Stephanides, 1948: pp. 19–20; pl. 7, figs. 57–60.

Moina rectirostris Gauthier, 1954: pp. 15–26; pl. 1, figs. *A–C*; pl. 2, figs. *A–D*; pl. 3, figs. *A–C*; pl. 4, figs. *A–C*; pl. 5, figs. *A–F*; pl. 6, figs. *A–D*; pl. 7, figs. *A–E*; pl. 8, figs. *A–B*; pl. 9, figs. *A–B*; pl. 10, figs. *A–C*; pl. 11, figs. *A–B*; figs. 2, *A–C*.

Moina rectirostris Scourfield and Harding, 1958: p. 25; fig. 7.

Moina rectirostris Šrámek-Hušek, Straškraba, and Brtek, 1962: pp. 247–249; figs. 89a–g.

TAXONOMIC NOTES

The taxonomic confusion involving *Moina brachiata*, seen in the above synonymy, has been in part due to the morphological variability associated with all species of the genus *Moina*. But for *M. brachiata* the error of the taxonomists has perhaps been more significant and has resulted in the species concept being so misconstrued that what is now described as *Moina brachiata* in Europe represents a fusion of the characters of two species, one misidentified.

The species was first described from Switzerland by Jurine (1820) as *Monoculus brachiatus* and was referred to as "Le Monocle à gros bras."

Au premier apercu on pourrait prendre ce monocle pour le *Daphnia quadrangula* de Müller, mais en considérant sa grosseur, le volume de ses bras, la longueur de ses barbillons et sa coquille lisse, on se convaincra que ce n'est pas la même espèce.

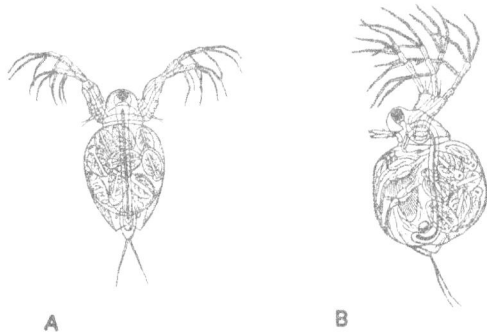

FIG. 2. Copy of Jurine's illustrations of *Monoculus brachiatus* (after Jurine, 1820; illus. by W. Vars).

Cet animal a comme le précédent deux manières de nager, horizontalement et verticalement; cette dernière est cependant la plus ordinaire; à chaque coup de ses bras vigoureux il s'élève bien plus que le *pulex*.

L'extension que prend la matrice pleine d'œufs est remarquable; elle donne à l'individu une forme presque carrée, d'autant plus que la coquille est tronquée postérieurement. Les œufs sont assez transparens pour permettre qu'on voie, au travers de l'enveloppe, les petits dont les chairs sont d'un rose jaunâtre.

Je vais extraire de mon Journal quelques observations relatives à la formation de la selle, dans cette espèce.

Le 27 Septembre, je rapportai plusieurs monocles à gros bras; quelques-uns avaient dans la matrice des œufs d'un blanc-jaunâtre, tandis que d'autres présentaient dans l'ovaire gauche seulement une matière rouge qui n'était pas ordinaire. J'isolai de ces derniers; le lendemain trois d'entr'eux muèrent et parurent avec une selle presque transparente; elle était composée d'un réseau à mailles hexagonales, dont le milieu était lisse, et les deux boules ovales relevées en bosse. Chez deux de ces individus la matière rouge de l'ovaire avait passé dans l'une de ces boules et la colorait fortement, tandis que l'autre restait vide. J'examinai ce qui était contenu dans cette boule rouge, et je distinguai une grande quantité de petits grains, semblables à la poussière des étamines de fleurs et enveloppés dans une gélatine transparente. Chez un autre, je vis la matière rouge, destinée à la formation de la selle, occuper l'ovaire gauche; dans le droit se trouvait celle des œufs laquelle était d'une couleur bien différente. Ce fait servira à expliquer comment les pontes peuvent alterner quelquefois avec les selles, et pourquoi les œufs se trouvent parfois mêlés dans matrice avec la matière de la selle." (Jurine, 1820: pp. 131–132.)

Jurine's comments on the external morphology were brief, and he failed to note any of the important species characters. He did designate body length and also gave very good illustrations of the species. The description and illustrations suggest an animal with these characteristics:

1. Length 7/12 of a line (Parisian line = 2.26 mm., Hutchinson, 1940: p. 372) or about 1.3 mm.
2. Large second antennae ("gros bras").
3. Head with a supraocular depression.
4. Head sharply separated from rest of body.

The latter part of his description was concerned with morphological changes that took place on a few specimens and must have referred to the formation of the ephippium. In describing the ephippium, he speaks of two spheres embossed on the shell; but he clearly states that only one egg formed and filled one of the spheres. In *Moina* the presence of a one- or a two-egg ephippium is a significant taxonomic character, and therefore the true meaning of Jurine's comments are of importance.

There are only three species of *Moina* now found in Central Europe; each one is quite distinct so that it is not too difficult to determine precisely which animal Jurine collected. Only two of these species measures 1.3 mm. or larger, and only one of these two has a supraocular depression; this is the form now identified in Europe as *Moina rectirostris* (Lilljeborg, 1900;

Keilhack, 1909; Wagler, 1937; Šrámek-Hušek, Straskraba and Brtek, 1962). Moreover, the ephippium of this form has two spheres embossed on the shell, one on either side, and contains only one egg. It is the only species with an ephippium that may be described as having spheres embossed on the shell. This species also has a large pecten on the claw which is an important identifying character.

This species is, in my opinion, the form Jurine had before him when he described *Monoculus brachiatus*. What he described as *Monoculus* (= *Moina*) *rectirostris* is a distinct form now designated as *Moina micrura* (see below pp. 28–29).

The chaotic condition that has followed Jurine's description of the species is a result not only of the inadequate original description but also of the misinterpretation of Jurine's comments about the number of eggs in the ephippium. Each subsequent description of Jurine's two species seems to have added to, rather than subtracted from, the confusion. In this regard it is necessary to consider also the history of *Moina rectirostris*. The pertinent references are these:

1. Lieven (1848) in his monograph on entomostraca, the first major work published after Jurine (1820) that referred to species of *Moina* correctly described and illustrated the present species as *Moina brachiata*.

2. Baird (1850) described the genus *Moina* for Jurine's two species *rectirostris* and *brachiata*. In so doing he illustrated a parthenogenetic female with a supraocular depression on the head that he called *Moiua brachiata* and an ephippial female and male both also with a supraocular depression and designated as *Moina rectirostris*. The parthenogenetic female as described and illustrated displayed few characters useful for specific identification other than the general body form and the supraocular depression. It is apparent, however, that it is the same form described by Jurine (1820) as *brachiata*. The sexual female likewise was inadequately described but the illustration of the ephippium is sufficient clearly to identify the form as being merely a sexual female of Jurine's *brachiata*. Baird could not have collected *rectirostris sensu* Jurine because this species does not occur in England. There is only one species found in England that has a supraocular depression. Baird may have been confused by the smaller size of the sexual female —the sexual females of *Moina* are frequently smaller than the parthenogenetic females.

3. Lilljeborg (1853), in determining that Jurine's name *Monoculus rectirostris* was invalid (see below, p. 28) concluded that *rectirostris* and *brachiata* were the same species.

4. Leydig (1860) disagreed with Lilljeborg (1853) that *brachiata* and *rectirostris* could be identical (but ignored Lilljeborg's comments about the validity of the name *rectirostris*) and described two separate forms from Germany. He properly described *brachiata* in

the sense of Jurine, although he exaggerated the size of the eye (*brachiata* has subsequently been illustrated with a very large eye); but he too was apparently confused by the variation in size and form that often occurs within the genus *Moina* and appears to have designated the same species as also being *Moina rectirostris*. He mentioned, for instance, that there were fourteen lateral teeth on the postabdomen and a very large pecten on the claws of his *rectirostris*. There is only one species of *Moina* that has these characters, *Moina brachiata* (Jurine).

5. Hellich (1877), as Leydig (1860) and Baird (1850), also identified the same form as both *M. brachiata* and *M. rectirostris*. *Moina brachiata* was characterized by a well-developed claw pecten and nine to ten lateral teeth on the postabdomen. His *rectirostris* had a pecten not quite so large and there were twelve to fourteen teeth on the postabdomen. With twelve to fourteen teeth this could only refer again to *brachiata*.

6. Gruber and Weismann (1880) described in detail the large species of Moina with the one-egg ephippium as *Moina rectirostris* following Baird (1850). They described Jurine's species *brachiata* (although they themselves had no specimens) as quite similar to *rectirostris* but the ephippium of *rectirostris* had only one egg. They believed that even though Jurine had referred to a single egg in the ephippium of *brachiata*, the presence of "deux boules ovales" implied that two eggs should develop and therefore *brachiata* must have a two-egg ephippium. In the same paper they described the two-egg ephippium of *Moina paradoxa* (=*macrocopa*). Having demonstrated therefore that a two-egg ephippium did occur in the genus, they were convinced that *Moina brachiata* must also have a two-egg ephippium.

Dating from Gruber and Weismann (1880), we have a clear description of the two species as they are recognized now in the literature.

7. Thus, in 1900, Lilljeborg referred to the large *Moina* species with a supraocular depression as *rectirostris*. However, he suggested that because this name, as used by Jurine, was invalid, authorship of the species be changed to Leydig (1860), the first person to describe the species in detail and should thereafter be considered a valid species name. Therefore the name appears in all recent faunal lists as *Moina rectirostris* Leydig, 1860.

8. But Keilhack in 1914 suggested that the name *rectirostris* should be changed to *Moina lilljeborgi* Schoedler, 1877, as the first described synonym for the species.

9. Despite all these suggestions and conclusions, Sars, in 1916, stated in describing *Moina brachiata* from South Africa:

The present species has generally been recorded by recent authors under the name *M. rectirostris*; but in my opinion this name cannot properly be supported, as it not only is a very inappropriate one, but, moreover, depends on an erroneous identification of this form with Müller's *Daphnia rectirostris*, which in reality belongs to a very different genus (*Lathonura*). It is here recorded under the name assigned to this species by some of the earlier authors, and indeed I am of opinion that it in reality is identical with Jurine's *Monoculus brachiatus*. (Sars, 1916: pp. 321–322.)

The nomenclatorial problems have, however, persisted to the present. *Moina brachiata* as now described in faunal lists combines the characters given it by Leydig (1860) with the two-egg ephippium on the authority of Gruber and Weismann (1880). Keilhack (1909) described *Moina brachiata* as follows:

Uber dem sehr grossen Auge eine tiefe Bucht. Schalenfelderung deutlich ausgerprägt. Unterrand nur vorn mit Borsten bewehrt. An den Seiten des Hinterkörpers nur 8–10 gefiederte Zähne. Nebenkamm der Endkrallen aus 8–9 langen Borsten bestehend. ♂: Vorderfühler in der Mitte knieförmig gebogen, am Ende mit 4 krallenförmigen Borsten.
Farbe blassgrün, undurchsichtig, meist mit Schmutz bedeckt. Länge: ♀ 1,2–1,6 mm; ♂ 0,5–0,6 mm.
Im Ephippium 2 Eier.
In Pfutzen und Schlammlöchern.

All specimens that I have examined from European museums labeled as *Moina brachiata* are either parthenogenetic females of what is also called *Moina rectirostris*, or sexual females of *Moina macrocopa* with a two-egg ephippium.

The name *Moina rectirostris* Leydig, 1860, is now usually assigned to that form described by Jurine as *Monoculus brachiatus*, while the animal described for example by Keilhack (1914) as being *Moina brachiata* does not exist.

It should be clear that this situation cannot be allowed to continue. I think it necessary, therefore, to return to the original meaning of each species as given by Jurine (1820). In doing this the large *Moina* species with the supraocular depression and large claw pecten should be referred to as *Moina brachiata* and designated the type species of the genus, since it was the only properly identified form referred to by Baird (1850) in his description of the genus *Moina*.

DIAGNOSIS

Large form, measuring 1 to 1.6 mm. long. Head with a supraocular depression located above the moderate-sized eye. Antennules long and thin, covered with many short setae. Second antennae well developed. Shell rim with thirty-five to forty setae that extend two-thirds of length of ventral rim.

Postabdomen with nine to fourteen lateral feathered teeth and one long bident tooth. Claw with a large pecten of eleven to fourteen teeth.

Ephippium bright yellow; reticulated only around the edges and with central embossed sphere. One sexual egg in ephippium.

Male antennules bent at one-third distance from base and with five to six short, thin terminal hooks. First leg without an exopod.

DESCRIPTION

Female

The head is rather broad and lacks hairs on the dorsal margin but has a patch of hairs on the ventral side behind the antennules (fig. 3, A). The head is somewhat more rounded than in *micrura*, but the eye is not as large in proportion to the size of the head (compare with fig. 9, A–D of *micrura*). The eye is near the dorsal margin and is composed of many facets surrounding a small pigmented spot.

The supraocular depression is not well developed but is readily apparent.

The antennules are long and thin and have setae arranged around them in rings. These setae are longest near the base. The short sensory seta is located on the lateral-anterior margin of the antennule at a point one-third the distance from the head.

The second antennae are behind, and dorsal to, the antennules. The basipod is very large and has several rings of setae. The two sensory setae that originate at the base of this segment are short; neither as long as the segment. The segments of the rami also have rings of short setae, and each ramus has an inner row of long hairs. The exopod has a medial row of teeth on the three distal segments. These groups of teeth extend over the proximal half of each segment.

The shell may be rotund or rectangular depending on the presence and number of embryos in the brood pouch. It has a granular surface and distinct reticulations.

The ventral shell margin has a row of thirty-five to forty-one long setae that extends backward along the margin but on only the anterior two-thirds of the shell margin. These are followed by about twenty groups of shorter setae each group consists of five to eight setae that increase in size posteriorly. These groups extend to the posterior shell margin and are there replaced by a row of short setae that are not arranged in groups. There is a pair of curved hooks, one on each valve, located just ventral to the point of the shell junction. These shell hooks hold the abdominal setae.

The first trunk limb of the female is similar to those of most other species of *Moina* (fig. 3, C).

The postabdomen is large and rather long (fig. 3, B). Its dorsal half is ornamented with wavy rows of fine setae but lacks long hairs.

The distal conical half of the postabdomen is armed with a long bident tooth and nine to fourteen feathered teeth on either side. One form of the species, *Moina brachiata caucasica*, is reported to have sixteen lateral teeth (Schiklejew, 1930). The base of the bident tooth has a row of fine setae.

The claw has a large pecten of eleven to fourteen teeth. This is one of the most conspicuous features of *Moina brachiata*. The distal half of the claw has many short setae. The ventral base of the claw has a "Basaldorn" of five to eight thin teeth.

The sexual female is frequently smaller than the parthenogenetic female, and the head seems to be slightly broader although this character is rather subjective. The ephippium forms on the dorsal half of the shell and is reticulated only around the edges; the central area is clear and embossed forming an oval sphere that has its long axis lying horizontally. This ephippium is usually bright yellow and contains only one egg.

The parthenogenetic females are 1.0 to 1.5 mm. long while the ephippial females are seldom over 1.3 mm. long.

Male

The general form of the male is somewhat rectangular. The head is not as long as the shell. The eye fills the end of the head, and is composed of small crystalline lenses surrounding a large pigmented spot. The supraocular depression is pronounced.

The long antennules are bent about one-third the distance from their base and have two sensory setae originating near the bend. One seta is short and with a rather thick base and stems from the medial margin; the second seta is long and thin and is on the lateral margin. The distal end of the antennule has five or six long curved hooks that are grouped together; they are not spread apart as in *M. micrura*.

The second antennae are not as well developed as in the female but have the same form.

The oblong shell is distinctly reticulated. Its ventral margin has long setae on the anterior half and groups of short setae behind these as in the female.

The first leg is large and well developed (fig. 3, D). Its third segment carries a long movable hook that may be extended ventrally or folded against the anterior margin of the leg. The terminal segment has three long setae; one is curved giving the appearance of being an accessory hook. The other two setae are feathered and extend out laterally from the leg. The medial margin of the leg is covered with a blanket of short setae that point distally. These various setae undoubtedly aid in clasping the female during copulation.

The postabdomen of the male is like that of the female; the pattern of setae on the dorsal margin and the lateral teeth are arranged similarly. The claw also has the large, well-developed pecten that consists here of eleven to fourteen teeth.

The testes lie lateral to the intestine on either side and the vas deferens extends onto the postabdomen and apparently opens near the ventral margin just proximal to the lateral row of feathered teeth. How-

FIG. 3. *Moina brachiata* Jurine. *A*. Slaghille, Denmark. 4-IX-26 (collected by K. Berg). *B*. Postabdomen of *A*.
C. Female first leg, Vichy, France, 29-VII-1886 (collected by J. Richard). *D*. Male first leg, Vienna, Austria
(British Museum Collection No. 1900.3.29.41; collected by Dr. Koellich).

ever, the precise position of this opening has not been determined for *Moina brachiata*.

The spermatozoa are spherical cells with many rays or axons as may also be found on spermatozoa of *M. micrura*.

The males are .8 to 1.1 mm. long.

Collections of this species may be found in many European Museums labeled either as *Moina brachiata* or *rectirostris*. The British Museum (Natural History) has a large number of properly identified specimens.

DIFFERENTIAL DIAGNOSIS

Moina brachiata may easily be distinguished from other species of the genus because it is the only large species with both a very large claw pecten and a distinct supraocular depression. If ephippial females are present, the single egg, yellow ephippium with an embossed globe may readily identify the species.

The claw pecten, composed of eleven to fourteen teeth, is the largest pecten on any Old World species of *Moina*. The only species with a comparable-sized pecten is *Moina wierzejskii*, which is restricted to the New World. *Moina wierzejskii* can be distinguished from *M. brachiata* by its lack of a supraocular depression and by the two egg ephippium.

Moina australiensis is very similar to *brachiata* but lacks a claw pecten and has hairs completely covering the head and most of the shell.

The only Old World species that might be confused with *Moina brachiata* are *M. hartwigi* and *affinis*. These two species both have small claw pectens and have rows of long hairs on the dorsal margin of the postabdomen which are absent in *Moina brachiata*. The males of *hartwigi* have only three hooks on the antennule (*brachiata* males have five or six) and the postabdomen claw lacks a pecten. *Moina affinis* males have the sensory setae of the antennules located very near the base, and the first leg has a greatly reduced hook.

Moina micrura on account of its smaller size, different head shape, and absence of a large claw pecten, should not be confused with *M. brachiata*. The antennules of *M. micrura* males have three or four hooks on the distal end.

Forms

There are two described infraspecific taxa of this species, but only one appears to be distinct. *Moina lilljeborgi* described by Schoedler (1877) was thought to be unique because of the presence of reticulations on the shell. These were supposedly absent on the shells of the typical form. I have found no specimens of *brachiata* that lack reticulations on the shell so I do not regard this distinction to be significant.

Moina rectirostris form *caucasica* was described by Schiklejew (1930) from the Caucasus in the U. S. S. R. This form is undoubtedly the same as *M. brachiata* but

has up to sixteen lateral teeth on the postabdomen rather than a maximum of fourteen teeth. This perhaps is a distinct character that might be of subspecific significance.

DISTRIBUTION

Type locality: Ponds near Geneva, Switzerland.

Moina brachiata is found throughout most of the Old World but is completely absent from the New World (specimens from the United States identified by Birge as *brachiata* are all *Moina micrura*). It is widely distributed in north temperate regions where it has been reported from most of Continental Europe and England, Russia, and Mongolia (fig. 4). The species extends southward to Italy and Greece, and has been collected from North Africa as well as South Africa.

The following reports appear to be reliable even though the species may have been listed under a different species name:

Europe:
   England: Baird (1850); Scourfield and Harding (1958)
   Sweden: Lilljeborg (1900); Eriksson (1934)
   Germany: Gruber and Weismann (1880); Eylmann (1887)
   France: Richard (unpubl.)
   Czechoslovakia: Šrámek-Hušek (1940)
   Hungary: Daday (1888)
   Italy: Marchi (1913); Pirocchi (1940)
   Greece: Stephanides (1948)
   Denmark: Berg (1929)
   U.S.S.R.: Sars (1903a); Fischer (1851); Hudendorff (1876)
Far East:
   Mongolia: Sars (1903a)
Middle East:
   Iran: Löffler (1956)
Africa:
   Tunisia: Gurney (1909)
   Algeria: Gauthier (1954)
   Union of South Africa: Sars (1916); Rühe (1914)

Specimens Examined

Europe:
   Austria:
      Vienna (Birge Collection slides No. 28, 35–38; from Richard collections)
      Vienna, collected by Dr. Koellick (British Museum, Natural History, collection No. 1900.3.29.41)
   Denmark:
      Slaghille, 4-IX-26, collected by Kaj Berg
      Egebjerg (Sars Collection, no date given)
   France:
      Vichy, 29-II-1886, collected by Richard (Birge Collection, slides No. 141–143)

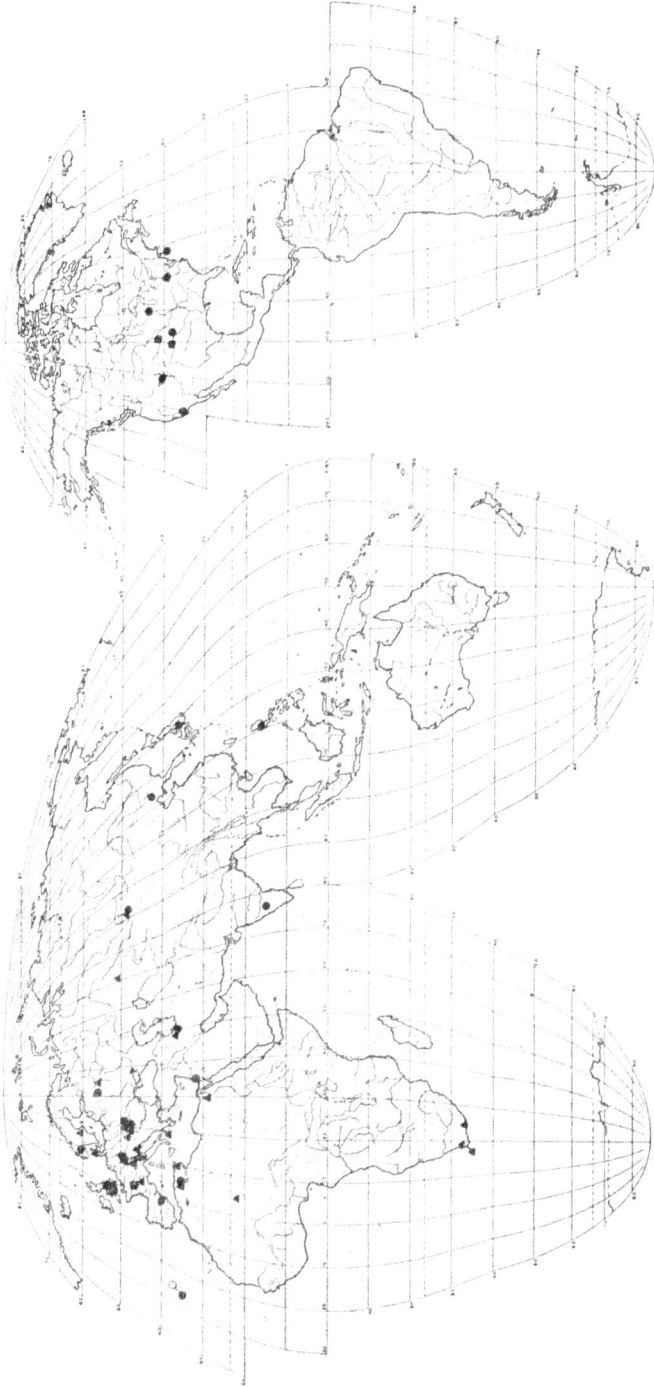

▲ Moina brachiata

● Moina macrocopa

Fig. 4. Geographical distribution of *Moina brachiata* and *macrocopa*.

Greece:
  Attica, collected by T. Stephanides (British
    Museum, Natural History)
England:
  Pond on Walton Common, N. Brampton (British
    Museum, N. H.)
Sweden:
  Upsala, collected by W. Lilljeborg (British
    Museum, N. H.)
Africa:
  Egypt:
    No locality given, (1938 British Museum, N. H.,
      collection No. 5.3.100–109)
  Union of South Africa:
    Cape of Good Hope, 1900 (Sars Collection)
    Plumstead, Nr. Capetown, 12-VII-03, determined
      by Rühe as *M. lilljeborgi* var. *salinarum*
      (Zoological Museum, Humboldt University,
      East Berlin collection No. 17463)

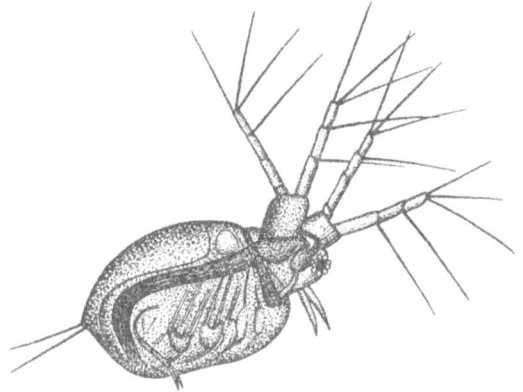

FIG. 5. Copy of Straus' illustration of *Daphnia macrocopus*
(after Straus, 1819; enlarged and illustrated by W. Vars).

MOINA MACROCOPA (STRAUS, 1820)

Daphnia macrocopus Straus, 1819: figs. 29–30.
Daphnia macrocopus Straus, 1820: p. 161.
Daphnia macrocopus Robin, 1872: pp. 452–465; pl. 16,
  figs. 1–5; pl. 17, figs. 2, 4, 5; pl. 18, figs. 3; pl. 19,
  figs. 1–2.
Moina flagellata Hudendorff, 1876: pp. 39–40.
Moina fischeri Hellich, 1877: pp. 55–56; fig. 22.
Moina paradoxa Weismann, 1877b: pp. 227–228; pl.
  10, figs. 36–45.
Moina paradoxa Gruber and Weismann, 1880: pp.
  82–92; pl. 3, figs. 1–2; pl. 4, figs. 6, 8, 9; pl. 5, figs.
  16, 18, 20; pl. 7, figs. 25–28.
Moina banffyi Daday, 1883.[4]
Moina paradoxa Eylmann, 1887: pp. 71–74; pl. 5,
  figs. 3.
Moina azorica Moniez, 1888.[4]
Moina banffyi Daday, 1888: pp. 112–113; pl. 3, figs.
  1–2.
Moina paradoxa Sars, 1890: p. 39.
Moina flagellata Matile, 1891: p. 130; pl. 4, figs. 19,
  19a.
Moina flagellata Birge, 1893: pp. 292–293; pl. 10,
  figs. 2, 4, 6, 9–11.
Moina banffyi Scourfield, 1903: pp. 437–438; pl. 24,
  figs. 5–8.
Moina macrocopa Sars, 1903a: p. 178; pl. 7, figs. 3,
  3a–c.
Moina rectirostris var. *Casañi* Arévalo, 1920: p. 166;
  3 figs.
Moina esau Brehm, 1936: pp. 289–290; figs. 1, A–F.
Moina nov. spec. Brehm, 1937b: pp. 22–23; figs. 1–3.
Moina esau var. *dschirofti* Hemsen, 1952: pp. 599–601;
  figs. 8–12.

---

[4] I have not been able to see the original descriptions for these
forms but have examined type specimens from Birge's collection.

Moina ganapatii Brehm, 1963: pp. 165–170; figs.
  11–15.

TAXONOMIC NOTES

Moina macrocopa was first illustrated by Straus in
1819 (fig. 5) and described in 1820 as *Daphnia macrocopus* (date of description used as date of species
name). His description was rather inconclusive and
this has led to some confusion. These are his
comments:

> Un millim. et demi. Tête sans bec. Deux longues
> antennes dans les deux sexes. Les rames très-grandes.
> Les filets du septième segment très-longs. Valves sans
> queue. La tige primitive des rames très-large. Le
> bouclier peu débordant. Les mamelons du sixième seg-
> ment presque nuls. Point de dentelures aux valves. Les
> valves séparées postérieurement sur le dos, dans le jeune
> âge. Une seule ampoule à l'ephippium. (Straus, 1820:
> p. 161.)

The major identifying character of his illustration
which can readily be identified as a species of *Moina*
is the absence of a supraocular depression on the head
(fig. 5). There are only three species of *Moina* found
in Central Europe, and of these, only one lacks the
supraocular depression. This is the form that is now
known as *Moina macrocopa*, so that, fortunately,
though many names have been given to this species,
the original designation can be unequivocally retained.
  Baird (1850) thought this species to be identical to
rectirostris sensu Jurine (1820); whereas Leydig (1860)
placed it in synonymy with *brachiata*. Robin (1872)
was the first to use the name *macrocopa* correctly.
  Largely owing to the indecisive nature of Straus'
(1819, 1820) description and illustration, to the force-
fulness of Baird and Leydig's opinions, and to the
obscureness of Robin's publication, as far as taxonomists were concerned, *macrocopa* was not recognized
as the proper name for this species until the early

1900's. As a consequence, prior to this time several names were applied. Even though the species was characterized in detail by Gruber and Weismann (1880), as *Moina paradoxa*, new synonyms continued to appear (see synonymy).

Straus (1820) credited Joblot (1754) with first describing this species. I have examined Joblot's manuscript and would hesitate to conclude this. Joblot's illustrations are too vague to assign either a specific or a generic name to them. However, this is not of concern for the validity of the species name because Joblot's paper was published prior to 1758 and he did not give his animal a new name.

DIAGNOSIS

Moina macrocopa is a large species; the parthenogenetic females are 1.0 to 1.5 mm. long while the males, .55 to .86 mm. long. The surface of the head and body is generally covered with long hairs.

The head is broadly rounded; a supraocular depression is usually absent. The eye is placed in the middle of the head and is of moderate size. The antennules are robust and densly ciliated. The second antennae are also densely ciliated and large.

The ventral shell rim carries a row of fifty-five to sixty-five long setae—these end at the posterior margin where they are replaced with a row of shorter setae, all of the same size.

The anterior seta on the penultimate segment of the female's first leg is toothed on its outer surface.

The postabdomen has seven to ten lateral feathered teeth and one short bident tooth. There is no distinct pecten on the claws.

The ephippium contains two eggs. The surface is completely covered with brown rectangular cells.

The male, as the female, is covered with hairs. The antennules are bent in the middle. There are five or six short, recurved hooks on the distal end.

The male first leg has a very large hook on the endopod and a long seta on the exopod. The spermatozoa are rod-shaped cells.

DESCRIPTION

Female

The body is oblong in lateral aspect with the head rounded in front. The head is not distinctly separated from the shell (fig. 6, *A–B*). The entire surface of the head and shell is normally covered with hairs. The hairs may occassionally be scarce, but I have examined no specimens where they were totally absent. The hairs are longer near the dorsal half of the head and shell, shorter towards the ventral margin. The hairs are absent from the dorsal half of the shell in ephippial females.

The head is broadly rounded, and there is little indication of a supraocular depression on most specimens. A slight supraocular depression is evident on

very large specimens (fig. 6, *B*). The eye is of a moderate size and is composed of a pigment spot surrounded by small crystalline bodies. The eye lies near the center of the head.

The antennules are large, robust, and armed with rings of setae. The setae are somewhat longer on the medial side. The sensory seta is very long and originates on the lateral margin at the mid-point of the antennule.

The second antennae are very stout. The segments are covered with many rows of setae. There are also medial rows of long, fine setae on both the exopod and the endopod rami. The two sensory setae at the base of the antenna are two-thirds as long as the basipod.

The shell has distinct reticulations formed by parallel lines. The hairs on the shell are arranged along these lines. The Old World subspecies of *macrocopa* has a shell with a granular appearance produced by many small pigmented bodies with a central body and many roots. The shell is smooth on specimens from North America.

The ventral shell rim carries a row of fifty-five to sixty-five setae that begins at the antero-ventral border and extends along the entire ventral margin. The setae are slightly longer on the front of the shell and become shorter posteriorly. Following this row, and extending along the posterior margin, are much shorter setae that may be grouped but are ungrouped towards the dorsal margin. In the North American subspecies, there may be three to four groups of rather large teeth following the ventral setae (fig. 6, *D*). These are then replaced by the smaller setae. In forms from the Old World these setae are rarely grouped (fig. 6, *E*). There are no large teeth along this margin in the Old World subspecies.

The first leg is very distinct in *Moina macrocopa* because the anterior seta on the penultimate segment has teeth on its ventral margin (fig. 6, *F*). These teeth are very stout and may be longer than the seta is wide. All other limbs are of the typical form and have the usual setae pattern.

The shell margin has a row of stout setae at the posterior-dorsal corner where the shell comes together. Dorsal to these is a pair of hooks, one on each valve. These hooks are not rounded as in *brachiata* but have a distinct angle.

The postabdomen is very broad and long; the distal, conical part is only one-fourth the total length (fig. 6, *C*). The dorsal margin of the postabdomen has rows of thin setae that slant towards the margin.

There are from seven to ten lateral feathered teeth and one short bident tooth on the postabdomen. The proximal arm of the bident is only as long as the feathered teeth while the distal arm is slightly longer.

The claw has one or two teeth on the ventral base. The dorsal surface of the claw has a row of fine setae that are slightly longer near the base. The proximal

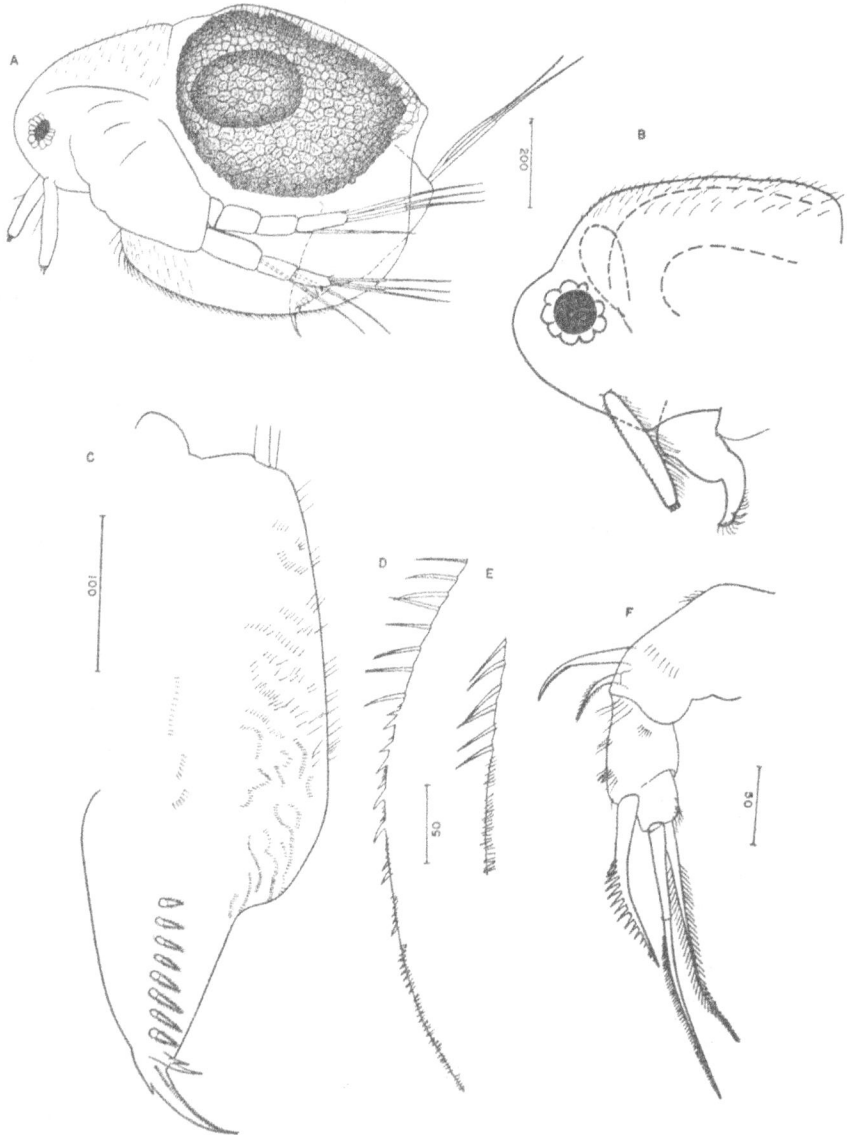

FIG. 6. *Moina macrocopa* (Straus). *A.* Sexual female, Hungaria Kup (?), 5-VI-1897 (from Daday Collections, labeled *Moina Banffyi*). *B.* Head of *Moina macrocopa americana* subsp. n. from roadside ditch near county road, one mile east of Cheyenne Bottoms Waterfowl Refuge, Barton County, Kansas, 26-VII-58. *C.* Postabdomen of *B.* *D.* Shell rim of *B.* *E.* Shell rim of *Moina macrocopa macrocopa* from Biskra, Algeria. Slide No. 145 of Birge Collection (identified as *Moina paradoxa* by Richard). *F.* First leg of female, lateral view, from roadside ditch near county road, one mile east of Cheyenne Bottoms Waterfowl Refuge, Barton County Kansas, 26-VII-58.

group of these setae may give the appearance of being a pecten, but usually a distinct pecten is not present.

There may be two or three short setae between the base of the bident tooth and the claw.

The ephippial female is generally smaller than the parthenogenetic female, only about 1.1 mm. long, but is of the same general form. The hairs which cover the body seem to be absent on the shell of the sexual females, or at least are absent from the area of the ephippium.

The ephippium itself contains two eggs. Straus (1820) found only a single egg in the ephippium. When these eggs first form, they lie horizontally and lateral to each other, however, and may appear as a single egg. They soon change their position, one moving behind the other, and rest with their vertical axes parallel to each other. This arrangement has been noticed by others, most notably Sars (1896) and Stephanides (1937) but Brehm apparently first recognized it in 1958 and described his *Moina lateralis* (=*belli*) on the strength of this as well as other characters.

The ephippium is brown and completely covered with polygonal or rectangular cells (fig. 6, *A*). The cells may be rather deep as seen on ephippia of the Old World subspecies or only superficial and flattened as in the North American form.

Male

The body of the male is densely covered with hairs (fig. 7, *C*). These are longer on the dorsal part of the shell than near the ventral part and are quite conspicuous along the dorsal margin itself. However, there are few hairs covering the head.

The head lacks the supraocular depression and has a large eye that fills the end of the head. The long antennules originate below the eye. The antennules are bent at the mid-point and have two sensory setae originating at or near the knee of this bend (fig. 7, *A*). One seta has a thick base and is on the medial margin while the second seta is longer and originates on the lateral margin of the antennule behind the other seta. The distal halves of the antennules are curved inward and have four to six short hooks at the tip. The hooks form a semicircle around a group of sensory papillae that project from the end of the antennule.

The head is separated from the shell by a groove which is more pronounced in the male than in the female.

The surface of the shell is reticulated and covered with hairs. The ventral rim has from thirty-five to forty marginal setae. The shorter setae along the posterior margin of the shell are either grouped as in the North American form or ungrouped and scattered as in the Old World subspecies. The postero-dorsal angle carries the pair of small hooks as in the female.

The abdominal setae may be seen to be inserted in these hooks.

The first leg of the male has a very large, recurved hook originating from the penultimate segment (fig. 7, *B*). The terminal segment carries three setae; the middle seta is long and hooklike. The other two setae are feathered. The penultimate segment is covered with many short hairs along the medial margin, and there is a seta arising from this surface opposite the hook.

The first leg has an exopod segment that terminates with a very long seta which is longer than the leg and reaches to the posterior margin of the shell. This seta is usually bent ventrally and projects well beyond the ventral margin.

The postabdomen of the male is similar in general form to that of the female; i.e., there are long hairs along the dorsal margin, and seven to ten lateral feathered teeth. The claw armature is also similar to that of the female. However, the conical part of the postabdomen is much broader, and the claw projects from the middle of the distal margin of the postabdomen (fig. 7, *D*). The two genital openings are just ventral to the claws; one on either side of the postabdomen. The claws are bent dorsally. The spermatozoa are rodshaped cells.

The males are .55 to .85 mm. long.

DIFFERENTIAL DIAGNOSIS

Moina macrocopa may be identified by the very broad head and the unique first leg. The anterior seta on the penultimate segment of the first leg has a series of teeth on the outer surface. The leg and the seta may be seen through the shell when the animal is placed under a low-power objective of the microscope.

The male of *macrocopa* may be identified by the antennules which are bent at the mid-point rather than near the head as in *australiensis*, *tenuicornis*, and *wierzejskii*. The exopod seta and the hook of the endopod of the first leg are very well developed and much larger than in these other species.

The males of *Moina belli* are very similar to *macrocopa* males, but the antennules of *belli* are bent about one-third the distance from the head; whereas in *macrocopa* they are bent in the middle. The first legs of the male and the spermatozoa of these two species appear identical.

The female of *belli* may be distinguished from the female of *macrocopa* by the first leg and by the smaller number of lateral teeth on the postabdomen.

SUBSPECIES

Moina macrocopa macrocopa (Straus, 1820). This is the typical form of the species and is restricted to the Old World. The general description of this form is as given above, but the following points should be used for identification: (1) the row of setae along the

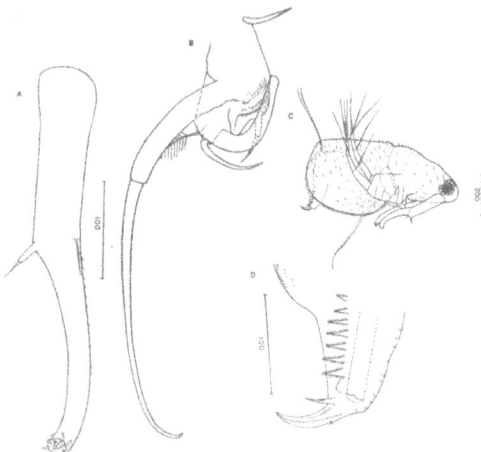

FIG. 7. *Moina macrocopa* (Straus). *A*. Male antennule. *B*. First leg of male. *C*. Mature male. *D*. Distal half of male postabdomen. *A–D* from roadside ditch along county road one mile east of Cheyenne Bottoms Waterfowl Refuge, Barton County, Kansas, 26-VII-58.

posterior shell rim are usually ungrouped and all of the same size; (2) the ephippium is ornamented with cubical cells that protrude above the surface of the shell.

Moina macrocopa americana subsp. n. The subspecies *americana* is restricted to North America and probably to the United States. I have found no records or specimens of *macrocopa* from Canada, Mexico, Central or South America. *Moina macrocopa americana* may be distinguished from the Old World *macrocopa* by the following characters: (1) the setae along the posterior shell rim are grouped, some being much larger than the others and there are usually two or three very large teeth located just posterior to the ventral row of setae (fig. 6, *D*); (2) the reticulated surface of the ephippium is composed of flat cells that are more trapezoidal than square in cross section.

Type specimens of this subspecies have been placed in the United States National Museum (Catalogue Numbers Holotype 123203; Paratypes 123204). This subspecies was originally collected from a roadside ditch along a county road one mile east of the Cheyenne Bottoms Waterfowl Refuge in Barton County, Kansas.

NOTES ON SYNONOMY

Moina flagellata Hudendorff, 1876, from a roadside ditch near a small village in Russia. In describing *flagellata*, Hudendorff observed that the entire ventral shell rim had many strong setae and that the armature of the first leg was very distinct. He stated that on the first foot there were two setae and that the convex side of one of these setae possessed a series of short

teeth: "an der convexen Seite mit einer Reihe spitzer Zähne besetzt, welche in der Mitte der Borste am grössten sind und sowohl gegen die Basis also gegen das Ende der Borste allmälig an Grösse abnehmen" (Hudendorff, 1876: p. 40). Both of these characters, but particularly the seta of the first leg, clearly identify *flagellata* as the same species as *macrocopa*. No type material is available for this species.

Moina paradoxa Weismann, 1877*b*. Weismann's (1877*b*) original description of this species was incomplete but a later description by Gruber and Weismann (1880) completely characterized the species. The illustrations of the parthenogenetic and sexual females and the male as given in the latter paper indicate that this species is identical to *macrocopa*. Furthermore, the illustration of the first leg by Gruber and Weismann (1880) should leave no doubt as to the identity of this form. There is no type material available.

Moina fischeri Hellich, 1877, from muddy waters in Bohemia (Czechoslovakia). *Moina fischeri* was described as having very sparse, fine and long hairs on the back of the head. The postabdomen was short and had only six to eight lateral feathered teeth and a rather short bident tooth. The claw of the postabdomen had fine teeth on the dorsal side but lacked a pecten. The ephippium was dark brown. There are only two species of *Moina* that fit this description, *belli* and *macrocopa*. The latter is the only one known to occur in Europe, and therefore *fischeri* must now be considered a synonym of *macrocopa*, and there is no reason to reject this synonymy, though no type material is avilable.

Moina banffyi Daday, 1883. I have examined species of this form in Daday's collections from Hungary and found all forms to be identical to *Moina macrocopa*. The primary feature by which it was distinguished from *macrocopa* was the presence of hairs on the head and shell. At the time *banffyi* was described, it had not been recognized that *macrocopa* also had hairs on the head and shell. Specimens identified by Daday are stored in the Termezettudomany Museum in Budapest, Hungary.

Moina azorica Moniez, 1888. A few co-types of *azorica* were found in the Birge Collection. The specimens were the typical *macrocopa* form. These specimens have been placed in the British Museum (Natural History).

Moina esau Brehm, 1936, Nilgiri Hills, India; collected by G. Evelyn Hutchinson. I have examined specimens of *esau* from the same collection used by Brehm (the collection is now in the Yale Peabody Museum) in his description of this species. There are no differences between this form and *macrocopa*. Specimens from the collection have been placed in the British Museum (Natural History).

Moina ganaptii Brehm, 1963, Yamuna River by Delhi, India. The description and illustrations of this species are identical to *macrocopa*, and it seems to be the same species. Brehm did not compare his specimens with *macrocopa* because he did not know that the latter species was also a "haired" *Moina*. No type material is known.

It seems rather unusual that there should be so many synonyms described for so characteristic a species as *Moina macrocopa*. This may be the result of two factors. First, the species was not well described initially. Second, the earlier descriptions of these various synonyms failed to mention the presence of the hairs that cover the head and shell of *macrocopa*. This can no doubt be blamed on the optics of the nineteenth century. Hellich (1877) was the first to recognize these hairs on his *Moina fischeri*, and Daday (1883) found them on the head of *banffyi*. It was not known that *macrocopa* also possessed such hairs until Scourfield and Harding (1941) concluded that what Scourfield (1903) had earlier called *banffyi* was actually *macrocopa*.

Brehm has repeatedly (1936, 1937a, 1938, 1958, and 1963) composed keys or discussed the differentiation of those species of *Moina* with hairs on the head but has not once included *macrocopa* with these "haired" moinids.

One other variable character of the species, the faint claw pecten, has also contributed to the confusion. Depending on the view of this setation on the claw, one might either conclude that a pecten is, or is not, present. This accounts for the different description of *banffyi* which was said to lack a pecten while *esau* was described as having one (Brehm, 1936). Both of these forms are definitely the same species.

DISTRIBUTION

Although *Moina macrocopa* has been reported from many parts of the tropical and subtropical regions of the world, it appears to be somewhat more restricted in its distribution than generally believed (fig. 4). The following reports seem to be valid:

Moina macrocopa americana:

North America:
  California: S. F. Light Collection
  Kansas: Goulden (unpubl. information)
  New York: Banta (1939)
  West Virginia: Goulden (unpubl. information; collected by S. Dodson)
  Wisconsin: Birge (1893)
  Idaho: Goulden (unpubl. information; coll. by B. and C. Durden)

Moina macrocopa macrocopa:

Europe:
  England: Scourfield and Harding (1958)

France: Pacaud (1952); Robin (1872)
Germany: Weismann (1877b)
Norway: Sars (1890)
Czechoslovakia: Hellich (1877)
Hungary: Daday (1888)
Spain: Arevalo (1920)
Azores:
  Moniez (1888)
Africa:
  Algeria: Blanchard and Richard (1890)
Middle East:
  Israel: Brehm (1937b)
  Iran: Hemsen (1952)
U. S. S. R.:
  Western Russia: Hudendorff (1876); Matile (1891)
Asia:
  Mongolia: Sars (1903a)
  Manchuria: Ueno (1937)
  Japan: Ueno (1927)
  India: Brehm (1936)
  Philippines: Tsi-Chung and Clemente (1954)

It is curious that *Moina macrocopa* is totally absent from Central and South America, most of Africa south of Algeria, the southern Pacific Islands and Australia. However, these areas are populated by other large species of *Moina*. We must presume that the other species are able to better exist under the conditions found there than *macrocopa*. In the southern United States, the West Indies, and South America, we find only *Moina wierzejskii*. In Africa, *Moina belli* is very common. In Australia and New Zealand, *Moina australiensis* and *tenuicornis* are the common large species.

Specimens Examined

M. m. americana:

North America:
  California:
    Roadside pond, Lancaster to Mohave highway, north central Los Angeles Co., 19-IV-37, collected by Dr. Nickelbacher (S. F. Light Collection No. 711)
  Kansas:
    Roadside pond, Barton Co., 10 miles NE Great Bend, 26-VII-58, collected by C. E. Goulden.
  Nebraska:
    No locality given, collected by Fordyce (Birge collection, slides No. 32, 34)
  Utah:
    Uintah Co., road 3 miles E. of Bonanza, alkali pool, 10-VIII-64, collected by B. and C. Durden.
  West Virginia:
    Small pond (locality unknown, collected by S. Dodson)

M. m. macrocopa:

Europe:
  England:
    Liverpool Canal, Leeds (British Museum, Natural History)
    Wales (British Museum, Natural History)
  Hungary:
    Collections from Buda (now part of Budapest), Kispest, Mohosti (?), and Nalacz (?), (Daday Collection; determined as *M. brachiata*, *M. macrocopa*, or *M. banfyii*)
Africa:
  Algeria:
    Biskra (Birge Collection no. 145—from Richard collections)
  Azores:
    Glade Tercure, 1887, M. Moniez det. as *M. azorica* (Birge Collection, slides No. 138–139)
India:
    Ootacomund, Ponds on Pykara Road, 13-XI-32, collected by G. Evelyn Hutchinson (det. by Brehm as *M. esau*)

MOINA MICRURA KURZ, 1874

Monoculus rectirostris Jurine, 1820: pp. 134–135; pl. 13, figs. 3–4.
Moina micrura Kurz, 1874: pp. 13–15; pl. 1, fig. 1.
Moina micrura Hellich, 1877: p. 56; fig. 23.
Moina propinqua Sars, 1885: pp. 29–35; pl. 5, figs. 4–5; pl. 6, figs. 1–5.
Moina micrura Matile, 1891: p. 129; pl. 4, figs. 18, 18a.
Moina weberi Richard, 1891: pp. 120–123; pl. 10, figs. 1–3.
Moina dubia de Guerne and Richard, 1892: pp. 527–530; figs. 1–2.
Moina paradoxa Stingelin, 1900: pp. 196–197; fig. 3.
Moina ciliata Daday, 1905: pp. 201–202; pl. 13, figs. 9–13.
Moina makrophthalma Stingelin, 1914: pp. 614–615; fig. 17.
Moina hartwigi Verestchagin, 1914?: pp. 13–14; figs. 6–7.
Moina micrura Birge, 1918: p. 704; fig. 1090.
Moina micrura Biraben, 1919: pp. 112–117; figs. 31–37.
Moina macrocopa var. *brevicaudata* Bär, 1924: pp. 102–104; pl. 6, figs. 4–5.
Moina dubia Gurney, 1927: pp. 66–67; fig. 5, *E–F*.
Moina dubia lacustris Rammner, 1931: pp. 623–634; figs. 1–4, 6.
Moina dubia macrocephala Rammner, 1933: pp. 362–366; figs. 5–8.
Moina latidens Brehm, 1933: pp. 684–685; fig. 11.
Moina dubia var. *parva* Jenkin, 1934: pp. 151–154; figs. 8, 8a; 9, 9a.
Moina dubia var. *baringoensis* Jenkin, 1934: pp. 155–160; figs. 10a–c, 11, 12, 12a–b.

Moina dubia Steuer, 1939: pp. 269–278; figs. 1–2.
Moina sp. Hemsen, 1952: pp. 597–598; figs. 6–7.
Moina cf. *weismanni* Brehm, 1953a; pp. 328–330; figs. 96–98.
Moina weismanni Tsi Chung and Clemente, 1954: pp. 106–107; pl. 2, figs. 6, 6a–d.
Moina micrura Šrámek-Hušek, Straskraba, and Brtek, 1962: pp. 249–252; pl. 90, figs. *A–L.*

TAXONOMIC NOTES

The systematic nature of this taxon has been greatly confused in the literature. In the absence of type material, it has been necessary, in order to identify this species, to rely on the early descriptions and illustrations, initially ignoring the present erroneous concept of this species.

The species was first described and illustrated by Jurine (1820) who thought it to be identical to Müller's (1785) *Daphnia* (= *Monoculus*) *rectirostris*. However, Lilljeborg (1853) has shown that Jurine was incorrect, that the species referred to by Müller actually belongs to the Family Macrothricidae and the genus *Lathonura* (the generic name was introduced by Lilljeborg, 1853). Jurine's species *rectirostris*, therefore, was founded on a misidentification and is an invalid taxon (see Article 49, International Code of Zoological Nomenclature, 1962).

In the present work it has been necessary first to determine what form Jurine had before him when making his description, and secondly, to ascertain the oldest available synonym to substitute for his name *rectirostris*. It should be mentioned here that Keilhack (1914) has previously suggested that the species name *rectirostris* be changed to *lilljeborgi* Schoedler, 1877. However, in the present study *lilljeborgi* is considered a synonym of *Moina brachiata* rather than of *rectirostris*.

According to Jurine, *M. rectirostris* was about 5/12 of a line in length (Parisian line = 2.26 mm., Hutchinson, 1940: p. 372) or about .95 mm. Unfortunately, his written description is inadequate and throws no light on what the species might be. The pertinent parts of his comments are these:

Cet auteur a bien observé que l'œil n'avait pas d'aréoles transparentes, et que la coquille était ciliée dans son bord inférieur. Lorsqu'il est question des petits, il dit: *Pulli duo in ovario matris similes, albidi, albumine hyalino cincti,* ce qui rend assez bien l'image de la couleur blanchâtre des œufs. Son silence sur l'œil de ces petits paraît extraordinaire, d'autant plus que, dans un autre paragraphe, il rapporte qu'ayant trouvé de ces monocles plus gros dont la coquille était opaque, ventrue, blanchâtre, et parsemée de quelques points noirs, il ajoute en parenthèse, (*Haec forte oculi pullorum haud tamen in ominbus aderant*); ce doute tenait vraisemblablement à la manière dont les petits étaient placés dans la matrice. (Jurine, 1820: p. 134.)

However inadequate Jurine's description is, his illustrations are excellent, considering the date that

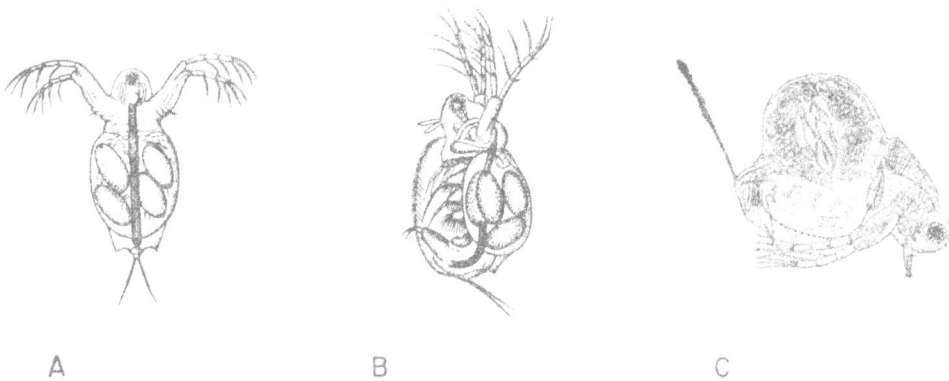

FIG. 8. *A* and *B*. Copy of Jurine's illustrations of *Monoculus rectirostris* (after Jurine, 1820; illustrated by W. Vars). *C*. Copy of Kurz's illustration of *Moina micrura* (after Kurz, 1874).

they were published, and are sufficient, with the measurement of body length, to identify clearly the form to which he must have referred. The illustration (fig. 8, *A* and *B*) shows a rather large eye in comparison to over-all body size, and a distinct supraocular depression.

Of the three species now found in Central Europe two are rather large species, measuring 1.1 to 1.6 mm. long, and neither is characterized by a very large eye. The third species may be as large as 1.2 mm. but more commonly is less than 1 mm. long. Furthermore it does have a large eye and a distinct supraocular depression. The shape of the head, which is characteristic for this latter form, is identical to that of the head illustrated by Jurine (compare fig. 8 with fig. 9, *A–D*). It seems certain, therefore, that this is the same form illustrated by Jurine (1820).

It is also apparent that, because this is the only small species of *Moina* found in Central Europe, it must be the same form described by Kurz (1874) from Austria as *Moina micrura*. Kurz's specimens measured about 1 mm. in length and were characterized by a large head. He illustrated (fig. 8, *C*) a large eye in proportion to the body (although he states that the eye is small), and the contour of the head is quite similar to that figured by Jurine.

I can only conclude from the published descriptions and from personal study of the species of *Moina* now found in Central Europe that Kurz's *Moina micrura* is the same as *Moina rectirostris* (Jurine, 1820). Since it is the first described synonym, *M. micrura* should be the valid name for this species.

DIAGNOSIS

A small species, .5 to 1.2 mm. long, but commonly .7 to .9 mm. The large head has a well-developed supraocular depression and a large eye. There are no hairs on either the head or shell. The ventral shell rim has eleven to twenty-five long setae followed on the posterior rim by groups of short setae. The postabdomen is short and narrow and has from three to eleven lateral feathered teeth and one very long bident tooth. The postabdomen claw lacks a strong pecten but has a row of thin setae that decrease in size distally. The ephippium contains one sexual egg. The ephippial shell may be ornamented with polygonal reticulations over its entire surface or these reticulations may be only around the edges and the shell indistinctly reticulated in the middle.

The male antennule has two sensory setae located one-third the distance from the head. The distal end of the antennule has three or four long hooks. The male's first leg has a well-developed hook. The spermatozoa are spherical and with many axons.

DESCRIPTION

Kurz (1874) described *Moina micrura* in the following manner:

Der Kopf ist verhältnissmässig sehr gross, während der Schwanz viel kürzer ist, als bei den beiden Arten. Die Länge des Thieres beträgt 1 Mm., der Schwanz (vom Grunde der Steüerborsten bis zur Klauenspitze) erreicht blos 1/3 der Körperlänge, während bei der borangehenden Art der Schwanz die halbe Körperlänge misst.

Der Kopfschild bildet blos ober den Armen eine schwache Leiste; vom Rostrum ist keine Spur vorhanden. Der Schalenvorderrand lässt die halbe Mandibel unbedeckt und hat ober derselben einen seichten Einschnitt, welcher an die gleiche Bildung bei *Sida* erinnert. Die sonstige Schalenbildung ist dieselbe, wie bei den bekannten Arten. Die Tastantennen sind kurz, bedeutend kürzer als bei *M. brachiata* und haben eine spindelförmige Gestalt; am Aussenrande tragen sie in der Hälfte ihrer Länge ein Tasthaar, nach hinten sind sie fast bis zur Spitze dicht und lang, aber fein behaart; am freien Ende sitzt das Büschel der kurzen Riechstäbchen. Die Ruderantennen sind sehr entwickelt; die beiden Tasthaare am Grunde des Stammgliedes sind sehr lang und fein gefiedert, ebenso das terminale Tasthaar vor den beiden Aesten. Diese sind normal gestaltet; ebenso Lippe und Füsse.

Der kurze Schwanz ist an seiner Basis dick, verjüngt sich aber gegen das Ende sehr stark; sein Bauchrand ist fast gerade, aber der Dorsalrand bildet dort, wo der After liegt, eine ziemlich starke Hervorquellung. Die Schwanzklauen sind sehr klein, ohne secundäre Bewaffnung. Von ihnen zieht aufwärts eine kurze Reihe von 6 starken, geraden Dornen, von denen der unterste ein Doppeldorn ist; die übrigen besitzen fein gezähnte Kanten. Die beiden Steüerborsten sind von enormer Länge, sie erreichen 2/3 der Körperlänge und sind somit etwa doppelt so lang als der Schwanz; ihr Endglied ist zart zweizeilig gefiedert.

Der Brutraum wird vom Körper durch eine quere Hautfalte und von der Schale aus durch eien hufeisenförmige Leiste abgeschlossen. Die auf einmal zur Entwicklung gelangenden Embryonen sind sehr zahlreich und treiben den Brutraum halbkugelförmig aug.

Das Auge ist verhältnissmässig kleiner als bei den bekannten Arten und mit zahlreichen Krystallkegeln, aber wenig Pigment versehen. Drei Muskelpaare sind zu seiner Bewegung vorhanden. Das Nebenauge fehlt, doch ist der Gehirnzipfel vorhanden, auf welchem es zu sitzen pflegt; es geht hier von demselben ein zarter Faden zur Stirn. Das Augenganglion ist vom Gehirn abgesetzt und tritt mittelst eines Augennerven in das Auge ein. Vom Gehirn entspringt ein augsteigender Nackennerv, welcher in drei Aeste sich theilt und fein verzweight in der Haut endigt. Uber den zellenförmigen Köper, der sich in der Nackengrube an die stark verdickte Cuticula anlegt, bin ich nicht ins Reine gekommen.

Die Bildung des Ephippiums und das Männchen bleiben mir unbekannt. Ich fand dieses Thier blos an einer einzigen Stelle in einem Mühlteich bei Maleschau unweit Kuttenberg, zusammen mit *M. rectorostris*, doch war diese Art zahlreicher als jene. (Kurz, 1874: pp. 13–15.)

It will be evident that the following description does not completely correspond with Kurz's account. Although type material of *Moina micrura* is not available, there is only one species found in Central Europe that can possibly be *Moina micrura*. The following description, moreover, is based on specimens collected from all parts of the world and thus includes a composite of characters of all forms of the species.

Female

In lateral aspect, the head is extended in an anteroventral direction, and the large eye lies almost contiguous to this margin (fig. 9, *A–D*). The ventral and anterior margins of the head are evenly rounded, but the dorsal margin is interrupted by a well-developed supraocular depression. In most forms this depression is very pronounced, however, in poorly preserved specimens, it may not be apparent. The entire head is distinctly separated from the body by a groove behind the second antennae.

The antennules originate well behind the eye, just below the second antennae, and are sometimes positioned on a small knob. They are of the typical shape, rather long and thin but lack heavy setae as found on the antennules of *M. macrocopa*. There is a medial row of long hairs that are vertically arranged in pairs. The sensory seta is located about one-third

to one-half the distance from the head and extends from the anterior margin. This seta is quite long and would reach to the tip of the antennule if pointed downward. The sensory papillae may also be very long.

The second antennae are rather fragile in appearance as compared to the other species of this genus (fig. 10, *B* and *C*). The terminal ramal segments just reach to the mid-point of the body, and the terminal setae extend almost to the posterior margin. The heavy setation is almost lacking in some forms of *micrura* while other forms have many setae. The four-segmented ramus (exopod) does have the medial row of setae groups that extend from the second segment to midway onto the fourth segment. On the last segment these setae do not extend to the end of the segment but are restricted to the proximal two-thirds of the segment.

The two sensory setae at the base of the second antennae may be quite long or may be short. The subspecies *dubia* has very long sensory setae, but the typical form of the species has rather short setae.

The shell is either oblong or rotund in shape, dependent on the size of the brood pouch. The surface of the shell has a granular appearance and a faint but definite pattern of rectangular reticulations that slant towards the antero-ventral shell margin. The ventral margin of the shell carries a row of eleven to twenty-five long setae that begin at the anterior margin and end midway along the ventral margin. The number of setae varies with the size of the individual. These long setae are replaced behind by very short setae that continue onto the posterior margin where the shell comes together. These smaller setae are grouped (fig. 10, *A*). There is a pair of curved hooks, one on each valve, at the posterior margin where the shell joins. Occasionally there are large glandlike bodies evident near the ventral part of the shell. These are more evident on shells of *Moina affinis* (fig. 15, *C*).

The postabdomen is short and rather slender—the distal conical portion composes only one-fourth of the total length (fig. 11, *A–F*). The dorsal margin of the postabdomen has many groups of short setae that are arranged in wavy lines slanting distally towards the margin. The distal part of the postabdomen has from three to eleven feathered teeth. This number varies considerably but is directly related to body size. The number of teeth is relatively constant within any one population, however, and probably represents a regional adaptation to a specific kind of habitat. There is also a very long bident tooth that originates on the same level as the feathered teeth but is about twice the size of the latter. There is a row of fine setae at the base of this bident.

The claw is long and sharply curved towards the distal end (fig. 10, *D–F*). The ventral base of the claw has from four to seven sharp teeth, the so-called

"Basaldorn." The dorsal side of the claw has a row of fine setae that extend from its base to the sharply curved part of the claw. The proximal group of setae are larger than the distal setae, but these usually grade into the latter. However, in some instances, as in the subspecies *dubia* and *ciliata*, these setae do not grade into the distal setae but are separated and give the appearance of being a distinct pecten (fig. 10, *E*).

The one-egg ephippium of the sexual female is reticulated around the margins and normally indistinctly reticulated in the middle (fig. 9, *E–F*). In the early stages of the development of the ephippium, the entire structure is heavily reticulated. The indistinct pattern of reticulations as found on the maturely developed ephippium must be due to the heavy sclerotization that occurs in the shell. These

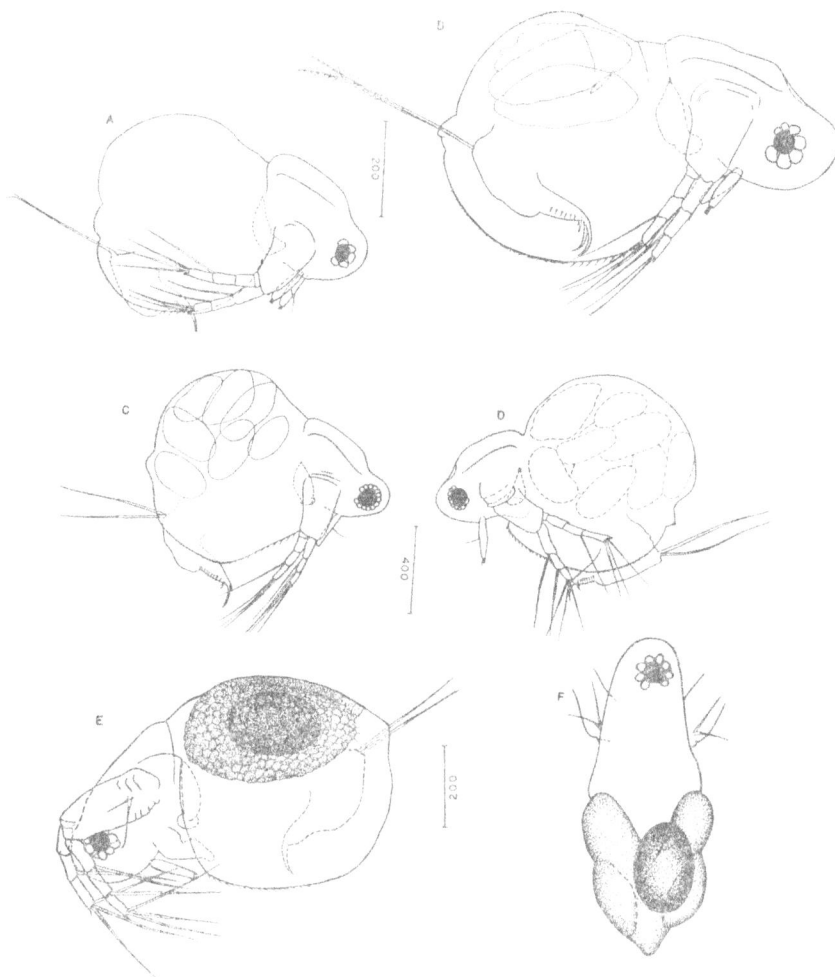

FIG. 9. *Moina micrura* Kurz. *A*. Carp pond east of Mallaha, Israel, 26-VI-63 (collected by U. M. Cowgill). *B*. Roadside pond three miles west of Emporia, Kansas, on Old Highway 50, 18-VI-65. *C*. Mare T. Vienp. et Briqueterie, Port au Prince, Haiti, 1-VI-1896 (Birge Collection No. 800 from Richard's Collections). *D*. Raised from Australian dried mud; precise locality and date not given (slide of *Moina propinqua*, Sars' Collection). *E*. Sexual female, lateral view, from roadside pool three miles west of Emporia, Kansas, 18-VI-65. *F*. Dorsal view of *E*.

reticulations are in a polygonal pattern, not circular. The central area of the ephippium is somewhat embossed above the remaining part of the ephippium.

Males

The males are more oblong in lateral view than the females, and the head is narrow and extended anteriorally (fig. 12, *A–B*). There is a well-developed supradeveloped supraocular depression behind the eye, rather than immediately above it as in the female, and the antennules originate below the eye rather than behind it (fig. 12, *B*). The eye is large and fills most of the anterior part of the head. The antennules are quite long and are bent at a point one-third the distance from the head. There are two sensory setae; one is short but quite thick at the base, and originates on the medial margin at the knee of the bend (fig. 12, *F*). The second seta is much longer and thinner and originates slightly below the first seta and somewhat lateral to it. The antennules are curved inward as in

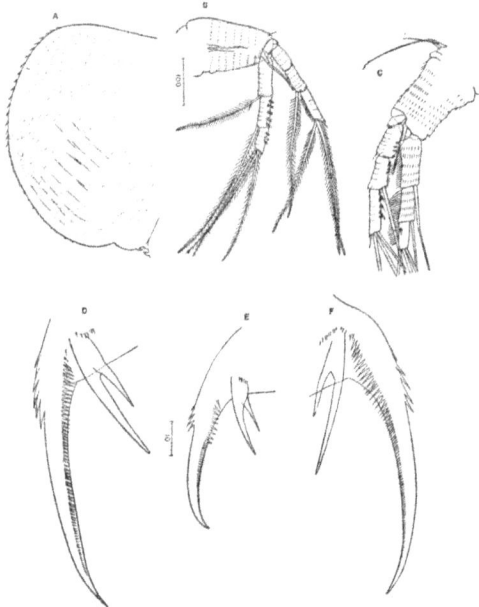

FIG. 11. *Moina micrura* Kurz. *A–F.* Female postabdomens. *A.* Mare T. Vienp. et Briqueterie, Port au Prince, Haiti, 1-V-1896 (collection No. 800 in Birge Collection). *B.* Roadside pond three miles west of Emporia, Kansas, on Old Highway 50, 18-VI-65. *C.* Barberspan, Southwest Transvaal, Union of South Africa, 6-IV-28 (collected by G. E. Hutchinson). *D.* Nature Preserve, Israel, 12-VII-63 (collected by U. M. Cowgill). *E.* Paraguay, no locality or date given (Daday type material of *Moina ciliata*). *F.* Fontaine du Hann, Rufisque, Senegal, 8-V-1890 (type material of *Moina dubia*, from Birge Collection, No. 716).

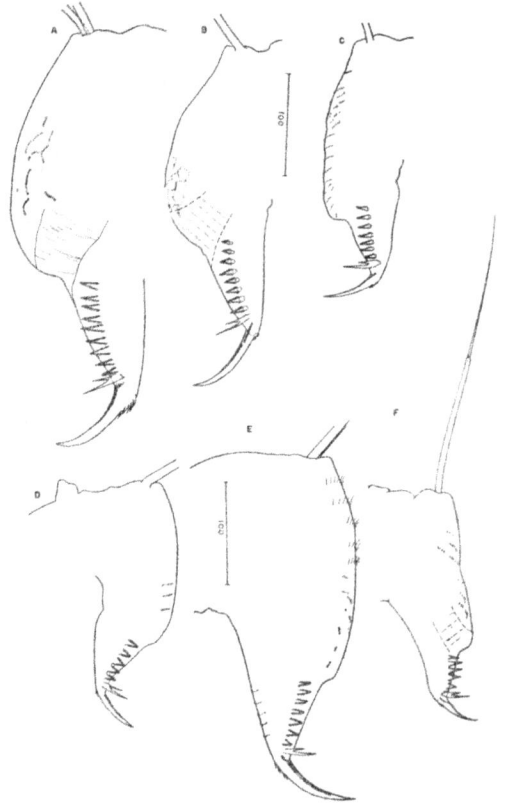

FIG. 10. *Moina micrura* Kurz. *A.* Female shell, roadside pond three miles west of Emporia, Kansas, on Old Highway 50, 18-VI-65. *B.* Posterior view of second antenna of *A*. *C.* Ventral view of second antenna of female from Mare T. Vienp. et Briqueterie, Port au Prince, Haiti, 1-VI-1896 (Birge Collection No. 800 from Richard's collections). *D.* Postabdominal claw of *A*. *E.* Postabdominal claw of female from Fontaine du Hann, Rufisque, Senegal, 8-V-1890 (type material of *Moina dubia*, from Birge Collection, No. 716). *F.* Postabdominal claw of female from Lewa, Usumbara, Burundi, Africa, 29-IX-1888 (collected by Stuhlmann, from collection in Zoologisches Muzeum, Hamburg University).

the other species and carry three to four hooks on the tip; the number of hooks is dependent on the age and the geographical race of the species. Most males that I have examined from North America have four hooks, occasionally three, and only the immature forms have less than three. Stephanides' (1948) specimens from Greece also had four hooks (this form was first identified as *M. salinarum* but Stephanides now believes it to be *M. dubia;* personal communication). All males of *dubia* described by Gauthier (1954) from Africa (Senegal) had three hooks as does also Sars' specimens of *M. propinqua* from Australia. However, there is no reason to believe that this must be a constant character. The number of hooks also varies in *M. brachiata, macrocopa,* and *wierzejskii.*

The first leg of the male carries a well-developed hook that may either be extended at a right-angle to the leg (fig. 12, *D*), or may be folded towards the leg (fig. 12, *E*). There is no exopod on the leg. The terminal segment has a long, slender seta that is straight in immature specimens and curved inward in mature specimens. There are also two feathered setae on this segment.

The postabdomen is similar to that of the female. The setae on the claw have the same pattern, and the number of lateral feathered teeth on the postabdomen vary as in the female. The genital opening appears to be on the ventral side of the postabdomen and proximal to the claw base (fig. 12, *C*). However, the opening is very difficult to find in this species unless the male is completely mature. In specimens that are noticeably undergoing a molt it is almost impossible to see. The spermatozoa are spherical cells that have

many radiating axons (fig. 12, *C*). In this regard, *M. micrura* is similar to *brachiata* and *hartwigi.*

The size of the animals varies considerably; the smallest form that I have measured, an adult female with embryos, was .47 mm. long. This specimen was from Israel. All adults in this collection were less than .6 mm. long. On the other hand, specimens of Sars' species *propinqua* from Australia measured as large as 1.2 mm. However, most forms of this species are from .7 to .9 mm. in length. The length of the body also varies considerably within each subspecies of *micrura.*

The size of the animal seems to be in part dependent on the habitat and on the availability of food. The size difference is not sufficiently constant to be considered a species character. Mature animals from the same collection may vary .2 mm. and sometimes as much as .4 mm. This great variation seems to be a character common to all species of *Moina.*

The males likewise show a considerable size variation. The smallest males are slightly less than .5 mm. while the largest are up to .77 mm. long.

Collections of this species have been deposited in the United States National Museum.

DIFFERENTIAL DIAGNOSIS

Moina micrura is at once distinguished from many members of the genus by its comparitively small size and by the complete absence of hairs on both the head and shell. It may be distinguished from *M. brachiata* by the absence of the large claw pecten and by the different type of ephippium. There are five species with which it may possibly be confused: *affinis, flexuosa, hartwigi, minuta, weismanni.*

Moina affinis is about the same size and form as *micrura* but can be distinguished by the presence of hairs on the head and shell; these are completely lacking in *micrura.* In addition, the ephippium of *affinis* has a pattern of round cells while in *micrura* the cells are polygonal. The male antennule of *affinis* has the two sensory setae located quite close to the head while in *micrura* they are at least one-third the distance from the head.

Moina micrura may be distinguished from *flexuosa* by the very different shape of the head and by the absence of the flexuous bend in the ventral margin of the shell. The ephippium of *flexuosa* is ornamented with many small knoblike protuberances, and the male antennule originates near the tip of the head rather than farther back as in *micrura.*

The females of *M. weismanni* have very short antennules that are spread apart and the front part of the shell is covered with fine hairs. The male's antennules are bent nearer the head than in *micrura* and have four short hooks on the distal end. The male first leg has a small terminal hook; whereas the hook on the first leg of *micrura* is quite large.

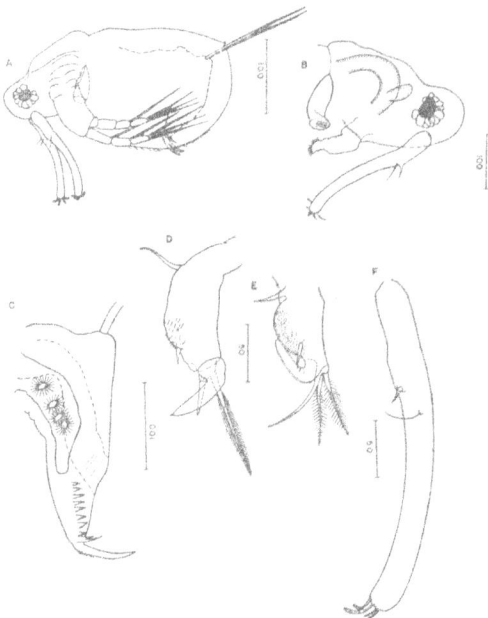

FIG. 12. *Moina micrura* Kurz. *A.* Mature male from roadside pond three miles west of Emporia, Kansas, on Old Highway 50, 18-VI-65. *B.* Head of *A.* *C.* Male postabdomen with spermatozoa, from Mare T. Vienp. et Briqueterie, Port au Prince, Haiti, 1-V-1896 (collection No. 800 from Birge Collection). *D.* Male first leg, from roadside pond one mile north of Lake Winnemucco, Humboldt County, Nevada, 4-VII-35 (collected by S. F. Light, No. 193). *E.* First leg of *A.* *F.* Male (immature) antennule, from Mare T. Vienp. et Briqueterie, Port au Prince, Haiti, 1-V-1896 (Birge Collection No. 800).

Moina hartwigi has hairs on the head, although only behind the antennules. The claw pecten is larger than the pecten on *micrura* claws, and the postabdomen of *hartwigi* has long hairs on the dorsal margin.

Moina minuta has a distinctly differently shaped head that is more triangular in outline than rounded as the head of *micrura*. The first leg of *minuta* also lacks the anterior seta on the penultimate segment. This seta is well developed in most specimens of *micrura* although it is rather short in the subspecies *dubia*.

NOTES ON SYNONYMY

Moina propinqua Sars, 1885. Gracemere Lagoon, Queensland, Australia. Sars distinguished *micrura* from his species *propinqua* by "the terminal part of the tail being far less produced and armed with a much smaller number of lateral denticles" in *micrura* (Sars, 1885: p. 30). His species measured up to 1.2 mm. long and the postabdomen carried as many as nine lateral teeth. These characters overlap with the known variation in *Moina micrura*, as given in the above description. I have examined Sars' type material of *propinqua* that is in the collection of the Zoologisk Museum, Oslo, Norway. I can see no major difference between these specimens and the typical specimens of *micrura* from North America, Europe, Africa, or Asia. Sars' illustrations of the ephippium and descriptions of the male are similar in all respects to *micrura*. The one male specimen that I examined had only three hooks on the antennule but otherwise was identical to the males of North American forms.

This form was raised from dried mud collected from a large temporary lake. Sars mentions that the mud had been collected at the bottom from depths of five and ten feet below the water surface. It would seem then that the species in this lake should be considered a planktonic form.

Moina weberi Richard, 1891. Lakes Singkarah and Manindjau, Sumatra. Richard distinguished *weberi* from other species of this genus by the shape of the head and the character of the postabdomen. Both features are, as illustrated by Richard, identical to *M. micrura*. I have examined specimens from Richard's collections from both of these lakes and can find little difference from typical *micrura* specimens. The postabdomen may be slightly narrower, but this difference is insignificant as a species character.

Paratype material from these collections has been placed in the British Museum (Natural History).

Moina dubia de Guerne and Richard, 1892. Rufisque, Senegal. Two co-type specimens of this species were found in the slide collections of E. A. Birge, and the original plankton collection which included many specimens was found in the Birge collections. Unfortunately, neither males or ephippial females were collected. Therefore, we must rely on Gauthier's (1954) description of the ephippium and the males of this form.

It is apparent that *dubia* is quite similar to the typical *micrura* specimens. The postabdomen claw does have a distinct pecten of approximately fifteen teeth (fig. 10, *E*; fig. 11, *F*), and the two sensory setae at the base of the second antennae are long. However, I have seen a comparable pecten on claws of *micrura* from South Africa and South America (subspecies *ciliata* from South America) that have only short setae at the base of the second antennae. Specimens from Israel also have the *dubia* type pecten on the claws, but the sensory setae are only moderately long. I have examined many specimens from both the Old and the New World and have compared them carefully with the specimens of *dubia*. Thus far I have found no constant difference not found in some degree in other forms of *micrura*. The males as illustrated by Gauthier (1954) are identical to males of typical *micrura* except that there are only three hooks on the distal end of the antennule. *Moina propinqua* also has only three hooks and the subspecies *ciliata* may have three or four hooks. At present I feel that *dubia* must be only a geographical subspecies of *micrura*. It seems to be restricted to the arid climates of North Africa and does not occur as often believed, in East African lakes or in Europe. Paratypes have been placed in the British Museum (Natural History).

Moina ciliata Daday, 1905. Lagoons along the Paraguay River, Paraguay, South America. Daday distinguished his specimens from *micrura* by the lack of hairs on the first antennae, by the absence of a comb or pecten on the claws, and by the many hairs or cilia on the dorsal margin of the postabdomen. The specific name is determined by the latter character. I have examined Daday's type material of this species that is now stored in the Termezettudomanyi Museum in Budapest, Hungary. Contrary to Daday's statements, I found his specimens to have the fine hairs on the first antenna and a pecten on the claw just as in most forms of *micrura*. However, the postabdomen does have several long hairs on the dorsal margin that may quickly characterize this subspecies. There seems to be no reason to consider the form as being anything more than a subspecies.

This subspecies appears to be widely distributed in South America. Biraben's (1919) *Moina micrura* and Rammner's (1933) *Moina dubia macrocephala* also had hairs on the postabdomen and undoubtedly belong to the subspecies *ciliata*.

Moina makrophthalma Stingelin, 1914. Eastern Cordillera and mountains of Colombia. I have not been able to locate Stingelin's collections and so have not seen the type material of this species. I can see no major difference between his description and illustration and those of *micrura*. The large eye is also a characteristic of *micrura*. However, Stingelin mentioned that an ocellus was present between the eye and the antennules. This is curious because there are only two other species of moinid Cladocera that have an ocellus—*Moina reticulata* and *Moinodaphnia macleayi*. The ocellus of *makrophthalma* may be an artifact or else a character of a mutant strain restricted to South America.

Moina latidens Brehm, 1933. Danau Bratan, Bali. The chief characters by which Brehm distinguished this species were the presence of only a slight supra-ocular depression and an unusual bident tooth on the postabdomen. This tooth had a broad base with the two points arising from the two distal corners of the base. In addition, the first feathered tooth, proximal to the bident, was only half as long as the other teeth.

I have observed these same characters on specimens of *micrura* from other parts of the world so I do not believe them to be significant differences. This species was collected in the plankton of a Sunda Island Lake; *Moina micrura* (=*weberi* and *dubia*) has been widely reported as a plankter from these lakes (Richard, 1891; Rammner, 1931, 1937; Brehm, 1933). It is improbable that a distinct species of identical size and habitat requirements should exist simultaneously in the same geographical region. No type material is available for this form.

Moina micrura is undoubtedly one of the most variable species of the genus *Moina*. It is difficult to believe that all of the above forms can belong to the same species. I have examined specimens from all parts of the world, including the Sunda Islands, South and Central Africa, Israel, Greece, India, Taiwan, Australia, North America, Haiti, and South America. I can see no constant variation in the geographical races that may serve to separate these forms into distinct species. The characters of the ephippial females and of the males, which should be of use, are of no help in this species. Besides the claw pecten, the only consistent character found on the sexual female has been the extent of reticulations on the ephippium while on the male the number of hooks on the antennule varies. These differences occur in the same populations and seem to change with age. They cannot be associated with any of the geographical races or subspecies.

It is understandable how these minor differences in body morphology associated with spotty collecting could lead investigators to conclude that each form represents a distinct species. Once it is realized that this same morphotype is found throughout the world, one can only conclude that all are indeed a single species.

I am, therefore, convinced that all of these forms should be considered together as one species. Until detailed studies can be made on the ecology and physiology of the various forms, I see no alternative. Recognition of the many previously described species could only lead to continued confusion within the genus.

DISTRIBUTION

As most other species of the genus, *Moina micrura* is commonly associated with small, temporary water bodies found primarily in the semi-arid and arid regions of the world. However, unlike the others, *micrura* may also be found in the plankton of large freshwater lakes. It has been reported in the plankton of lakes in Germany (Rammner, 1931), Czecko-slovakia (Šrámek-Hušek, 1940), in the large Rift Valley lakes of Africa (Daday, 1910; Jenkin, 1934; Delachaux, 1917; Verestchagin, 1914), Sumatra (Richard, 1891), Java (Rammner, 1937), Bali (Brehm, 1933), and in the southern United States in Lake Pontchartrain, Louisiana (Birge collections). Furthermore, Worthington and Ricardo (1936) have found that this species has a daily vertical movement in the lower waters of some of the East African lakes.

It is probable that the planktonic mode of life has greatly added to the ubiquitous nature of this species, for *M. micrura* is the most widely distributed of all species of *Moina*. It is found in all parts of both the Old and the New World except for the cold-temperate regions (fig. 13). It has not been found north of the Great Lakes in North America, north of Germany in Europe, or north of Moscow in the Soviet Union. It has been reported throughout the Middle East, Asia, Africa, Eastern Australia, and most of South America. *Moina micrura* also occurs on the Caribbean Islands as well as on islands in the Pacific including the Philippines.

This species is the most ubiquitous and probably the most variable of all species of *Moina*.

Specimens Examined

North America:

Arkansas:

Chickasaba (Birge Collection, slide no. 27)

California:

Swamp, one mile S. of Upper Lake, Lake Co., 16-VII-36 (S. F. Light Collection No. 422)

Small water hole in dry creek in High Valley just over the hills from Clear Lake Oaks, on Clear Lake, 21-VII-36 (S. F. Light Collection No. 427)

Western border of Calaverio Co., large pond on Valley Springs, Stockton Rd., 20 miles E. Stockton, 14-V-51 (S. F. Light Collection No. 328)

Indiana:

Roadside pond, south of Bloomington, Monroe Co., 24-IX-60, collected by C. E. Goulden

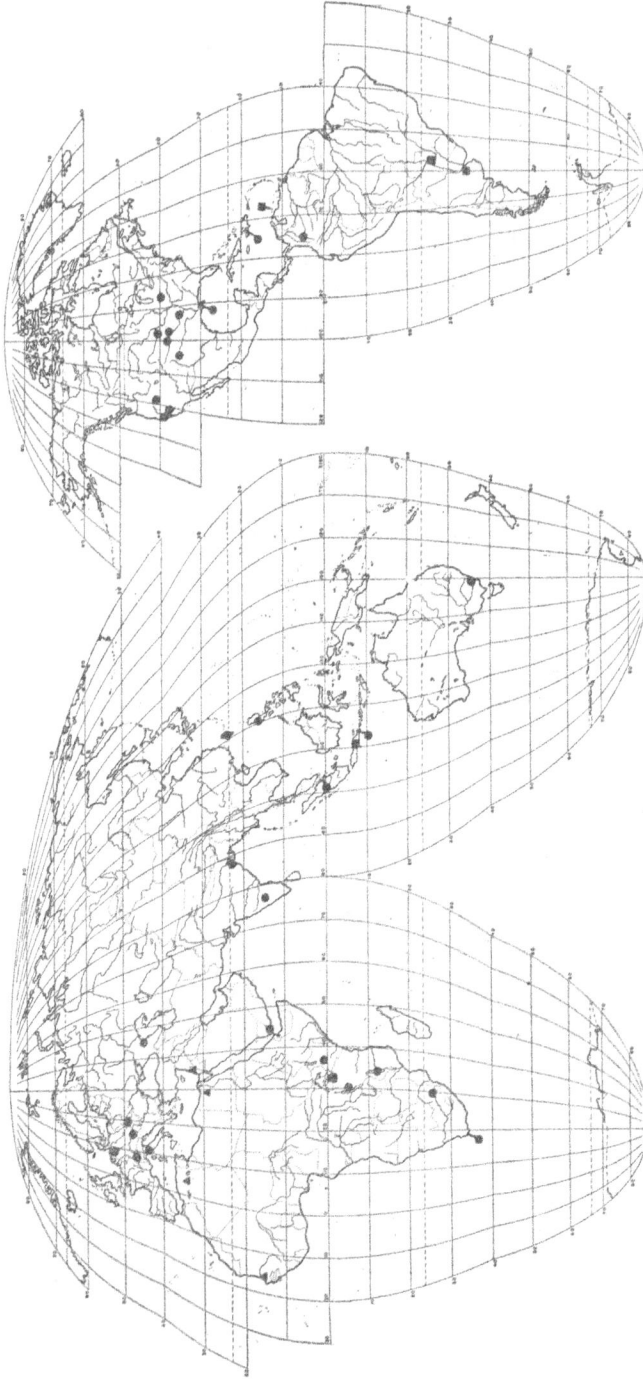

● Moina micrura typica
■ Moina micrura ciliata
▲ Moina micrura dubia

FIG. 13.   Geographical distribution of subspecies of *Moina micrura*.

Kansas:
Roadside ditch, Cheyenne Bottoms, Barton Co., 27-VI-58, collected by C. E. Goulden
Louisiana:
Lake Charles, 12-IX-06 (Birge Collection)
Nebraska:
Lincoln, 27-IX-1894 (Birge Collection, slides No. 18–23)
Nevada:
Small pond, one mile N. Winnemucca, Humboldt Co., 7-IV-35 (S. F. Light Collection No. 193)
Texas:
Lubbock, playas near town, collected by Vernon Proctor, 1-IX-64
Haiti:
Mare T. Vienp. et Briqueterie, Port au Prince, 1-V-1885 (Birge Collection No. 800, from Richard's material)
Mare Brinville, Port au Prince, April, 1895 (Birge Collection No. 752, from Richard's material)
Paraguay:
No locality or date given (Daday Collection, det. as M. ciliata)
Africa:
Algeria:
Lac du ous (?) (Sars Collection—no date given, No. Mp 141, F. 9286)
Kenya:
Lake Baringo (British Museum, Natural History, det. by P. M. Jenkins as var. baringoensis, No. 1939.4.25.13)
Senegal:
Fontaine de Hann, Rufisque 8-V-1890, collected by M. Chevreux (Birge Collection No. 716, from Richard's material)
Union of South Africa:
Princess Vlei, Cape Peninsula, 5-II-29, collected by G. Evelyn Hutchinson
Barberspan, Southwest Transvaal, 6-IV-28, collected by G. Evelyn Hutchinson
Tanganyika (Tanzania):
Lake Victoria (Sars Collection No. Mp 141, F9289)
Burundi:
Usumbura, 29-IX-1888, collected by F. Stuhlmann (Hamburg University, Zoologisches Museum)
Middle East:
Aden:
Middle tank near boat dock, 21-II-32, collected by G. Evelyn Hutchinson
Israel:
Carp pond east of Ein Mallaha 26-VI-63, collected by Ursula M. Cowgill
Nature Preserve, Lake Huleh 12-VII-63, collected by Ursula M. Cowgill

Far East and Southeast Asia:
India:
Calcutta, (Daday Collection, no date given)
Ootacoamund, small pool near milestone 4, 9-XI-32, collected by G. Evelyn Hutchinson
Sumatra:
Lake Singkarah, Peche pelagique, collected by Max Weber, 3-V-1888 (Birge Collection Nos. 702 and 703 from Richard's material det. as M. weberi)
Taiwan:
Ranch: Tam, plankton (collected by M. Tsukada, 12-X-64)
Australia:
Sydney, raised from dried mud (Sars Collection, no date or precise locality given, No. 53.2/11)
Europe:
Italy:
Caseina Corte Nuova, rice fields, Parma, 9-VI-61 (collected by A. Moroni)

MOINA AFFINIS BIRGE, 1893

Moina affinis Birge, 1893: pp. 290–292; pl. 10, figs. 1, 3, 5, 7, 8, 12, 13, 14.
Moina affinis Birge, 1918: p. 705; figs. 1092B, 1093B, 1094.
Moina irrasa Brehm, 1937a: pp. 95–96; figs. 9, 10a, 11.
Moina irrasa Brooks, 1959: p. 622; fig. 27.46.
Moina rectirostris Moroni, 1962: pp. 25–26, figs. 19–21.

DIAGNOSIS

Parthenogenetic females .8 to 1.2 mm. long. Head and shell covered with hairs. Head with a supraocular depression. Antennules short and with a long sensory seta. Exopod of second antenna with a vertical row of teeth that extends full length of each segment. Ventral shell rim with twenty to twenty-seven long setae followed behind with a row of short, ungrouped setae. Postabdomen with a long bident tooth and seven to fourteen feathered teeth. Claw with a pecten of ten to fourteen sharply pointed teeth. Male with sensory setae of antennules very near head. Distal end of antennules with four hooks. First leg of male with a small hook on the third segment, a long hooklike seta on the terminal segment, but without an exopod. Testes with spherical spermatozoa.

DESCRIPTION

This species was first collected and described by Birge (1893) from Wisconsin (fig. 14). Birge described the species as follows:

Female

The head closely resembles that of M. rectirostris, Jur., being somewhat rounded anteriorly, having a deep depression above the eye, and being without an angle on the ventral margin posterior to the antennules. As seen

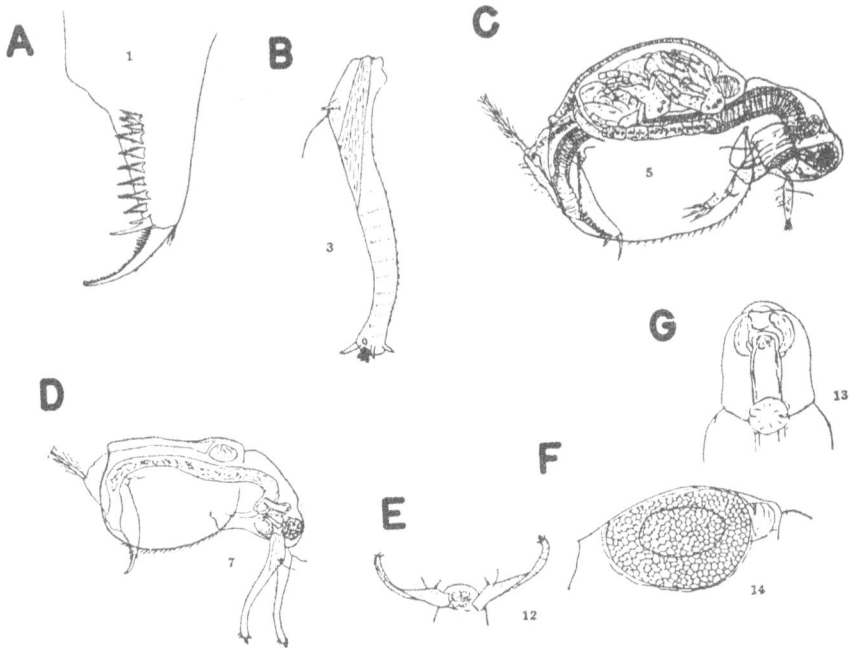

Fig. 14. Birge's original illustrations of *Moina affinis* (numbers refer to Birge's original figure numbers). *A.* Postabdomen. *B.* Male antennule. *C.* Parthenogenetic female. *D.* Male. *E.* Male head. *F.* Ephippium. *G.* Dorsal view of female head.

from above, the head is rather long and narrow and shows the supraocular depression very distinctly. The outline of the sides of the head is slightly concave in the middle and the sides round over evenly into the anterior margin. The valves are fringed on the margin with setae more closely set than in *M. rectirostris*, and are marked by transverse, anastomosing lines, giving an appearance to the shell somewhat like that of *Simocephalus*. These markings can be seen when the shell is examined uncovered and without water, and also, by careful manipulation, under a cover glass. These striae resemble those of *M. Lilljeborgii*, Schdl. as figured by Lilljeborg, ('53, p. 38, pl. II, f. 4f.), and still more closely those of *M. propinqua*, Sars, as described and figured by him, ('85, p. 31, pl. VI, f. 1.).

The structure of the legs agrees exactly with that of *M. rectirostris*, as described by Gruber and Weismann, ('77, p. 70–72).

The post-abdomen has a relatively long post-anal portion, which is armed with 9–11 serrate teeth and a bident longer than the adjacent tooth. The caudal claws have a pecten of 12–15 teeth at the base and are denticulate.

The ephippium contains one egg whose long axis is parallel to that of the body. The ephippium is densely reticulated over its entire surface.

The antennules are of moderate size, being apparently somewhat smaller than the figures of *M. rectirostris* would indicate for that species. The usual anterior sense-hair is placed a little proximal to the middle and its length is about one-half that of the antennule. The antennules are fringed on the posterior side by a dense growth of very fine hairs, visible only with a high power of the microscope, but easily disclosed by treatment with osmic acid.

The antennae resemble in general those of *M. propinqua*. Transparent, with sometimes a tinge of violet. Length, 0.8–1mm. Height, 0.4–0.5mm.

Male

The male is about 0.3–0.6mm. in length, and agrees in structure with the usual type of the males of this genus. The antennules are modified into powerful claspers. They are broad at the base in the antero–posterior direction and are inserted near the vertex, so that the head projects but little beyond them. They are geniculate, the angle occurring about 1-5 of the length from their insertion. At the bend are two sense hairs, one short and stout and the other long and slender. There are four hooks at the distal end of the antennule.

The first foot has a hook and is without a flagellum.

The spermatozoa are spherical or oval, and never have radiate projections of protoplasm. (Birge, 1893: pp. 290–292).

Female

The head of the female in lateral aspect is like *micrura*, but in *affinis* the head is covered with hairs (fig. 15, *A*). These hairs are very dense along the ventral and dorsal margins but are absent from the tip of the head.

There is a supraocular depression but it is not as well developed as in *micrura*.

The eye is large; the pigmented spot occupies a large area and is surrounded by large crystalline lenses.

The antennules are very short and originate behind the eye, just below the second antennae. The sensory seta is long and projects from the mid-point of the antennule on the medial side.

The second antennae are larger in respect to the body size than are the antennae of *micrura* and are much more pubescent. The two sensory setae at the base of the antennae are long; one being slightly longer than the basipod. The seta that arises between the two rami of the antennae is almost as long as the first segment of the three-segmented endopod.

There are many rings of setae on all segments of the antennae, and the four-segmented exopod has a long row of stout teeth that begins on the first segment and ends at the tip of the terminal segment (fig. 15, *B*). The first segment has two groups of these teeth that

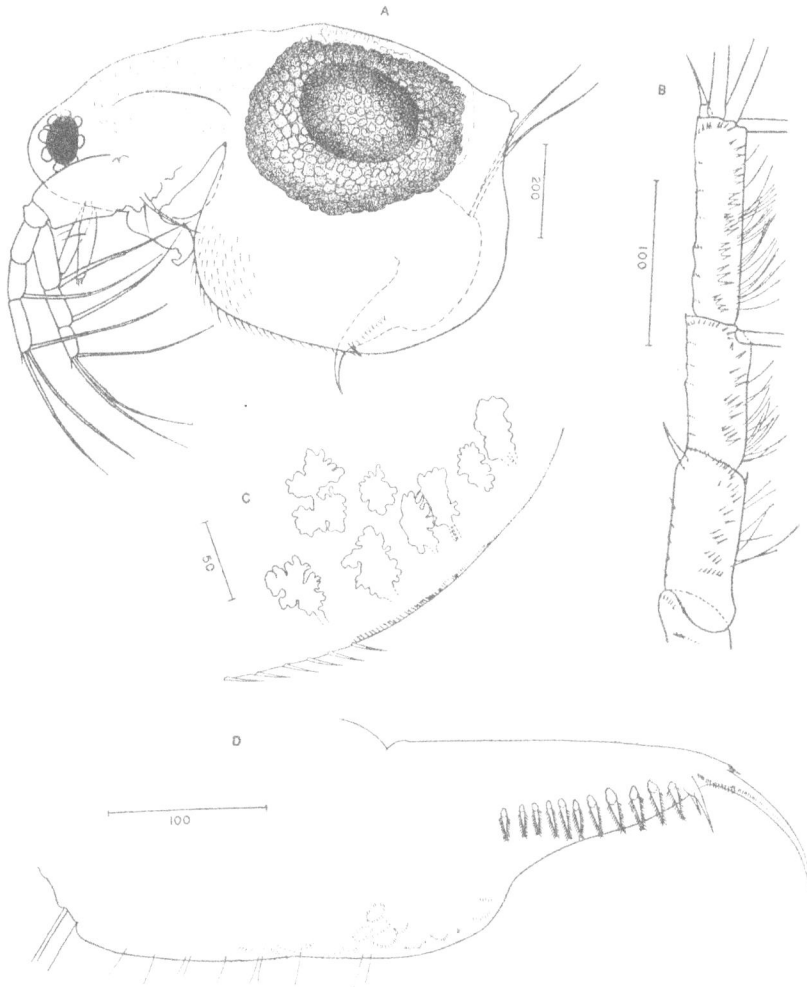

FIG. 15. *Moina affinis* Birge. *A*. Sexual female from roadside ditch pool five miles east of Strong City, Kansas, on Old Highway 50, 18-VI-65. *B*. Exopod of second antenna, cow pond approximately ten miles south of Yates Center, Kansas, Highway 75, July, 1959. *C*. Shell margin with long ventral and short posterior setae and with gland-like cells on shell, same as *B*. *D*. Postabdomen of female, roadside pond two miles east of Cheyenne Bottoms Waterfowl Refuge, Barton County, Kansas, 26-VII-58.

border the distal rim of that segment. The second segment has about five groups. These teeth lie horizontal to the axis of the antennal segment but slowly turn to parallel the axis at the distal end. There are five to six teeth in each group. The third segment of the exopod has four groups of teeth with four or five teeth in each group. The last segment has about seven groups of teeth.

These teeth are restricted to the medial side of the antennae. The number of groups and the number of teeth in each group may vary, but the general position of the teeth appears to vary little.

Both rami of the second antennae are also bordered with long thin hairs that form a vertical row on the inner margin of each ramus.

The surface of the shell is reticulated and has hairs that cover the anterior half of the shell. The hairs are particularly dense near the ventral margin. There may also be pigment cells bordering the ventral shell rim (fig. 15, C).

The ventral shell rim has a row of twenty to twenty-seven long setae that are replaced behind by a row of much shorter setae. These shorter setae are of equal size and extend from the latter third of the ventral shell margin onto and along the entire posterior margin. There is a pair of hooks at the dorsal end of the posterior shell rim.

The first leg of the female is similar to that of the other species of *Moina*. The anterior seta of the penultimate segment is finely striated and lacks fine hairs.

The postabdomen is long and slender (fig. 15, D). The distal conical portion comprises about one-third of the total length, and the two abdominal setae are at least one-half again as long as the postabdomen.

The dorsal part of the postabdomen has many short setae that are arranged in wavy rows that interconnect to form a webbed pattern. The dorsal margin itself has several long hairs that may be aligned in horizontal rows.

The distal part of the postabdomen may have from seven to fourteen lateral feathered teeth that are usually long and rather thin. The bident tooth is also very long; the distal arm is about twice the length of the proximal arm. There is a row of fine setae at the base of the bident.

The claw has a "Basaldorn" of four or five thin teeth. The dorsal base of the claw has a distinct pecten that consists of ten to fourteen sharply pointed teeth. The remaining part of the claw has a row of fine setae.

The adult parthenogenetic female is from .8 to 1.2 mm. long. Birge (1893) mentions the maximum size as 1.0 mm., but I have measured specimens that are 1.13 mm. long and have seen slightly larger ones. It is doubtful, however, that specimens larger than 1.2 mm. will be found.

The ephippium of the sexual female contains one egg and is completely reticulated with a pattern of round cells (fig. 15, A).

Male

The head of the male is rather oblong and has a supraocular depression above the eye—not behind the eye as in *micrura* (fig. 16, A). The eye is large and fills the tip of the head.

The antennules originate very near the tip of the head and are curved inward. The indistinct knee or bend is very near the head (fig. 16, C–D). There is a series of small protuberances along the medial part of the bend. There are two sensory setae, as is normal for this genus; one originates at the anterior margin and the second is located behind the first and very near the posterior margin. The latter seta is long and thin while the first seta is shorter and has a stout base. The anterior margin of the male antennule has many fine hairs that may also be seen on the medial margin. These hairs cover the proximal half of the antennule. The distal end of the antennule has four incurved hooks. I have seen one immature male with only two hooks so this number seems to vary with age as in *M. micrura*.

The head is only sparsely covered with hairs; but the anterior portion of the shell is densely covered with hairs, particularly near the ventral margin.

The ventral shell rim has twenty to twenty-two setae followed by a row of shorter, unorganized setae as in the female.

The first leg has a poorly developed hook that is curved inward (fig. 16, B). The terminal segment has three setae; two are feathered, the third seta is reduced to merely a small, knoblike protuberance. This third seta is much longer and hooklike in *micrura* and *brachiata*, but in *affinis* is considerably reduced in size.

The third segment of the male first leg is covered with fine setae on its medial margin. These setae are not as dense as in either *micrura* or *brachiata*. There is no exopod on the first leg, and the leg is poorly developed in comparison with either *brachiata* or *micrura*.

Specimens and collections of this species have been placed in the United States National Museum and the British Museum (Natural History).

Moina affinis may be distinguished by its small size (maximum length about 1.2 mm.), supraocular depression on the head, and the presence of hairs on both the head and shell. These hairs, and the ungrouped setae along the posterior shell rim, serve to differentiate *affinis* from *micrura* and *brachiata*. In the latter two species, the setae on the posterior shell rim are grouped. The species with which it might be confused are *wierzejskii* and *weismanni*. *Moina wierzejskii* has a two-egg ephippium that is ornamented with knoblike protuberant cells, whereas *affinis* has a one-egg ephippium that has a pattern of large round cells. *Moina wierzejskii* has a very broad head that only infrequently may have a supraocular depression. *Moina affinis* has a much smaller and a narrower head that always has a well-developed depression above the eye. The antennules of *affinis* are short; whereas the antennules of *wierzejskii* are long and very robust.

The males of *affinis* and *wierzejskii* are very similar; the antennules of both species originate at the front end of the head, and the sensory setae of the antennules are located very near the head. However, the tip of the antennules of *wierzejskii* usually have a single hook that points forward in addition to the medially directed hooks. There is also a difference in size between males of the two species; *wierzejskii* males

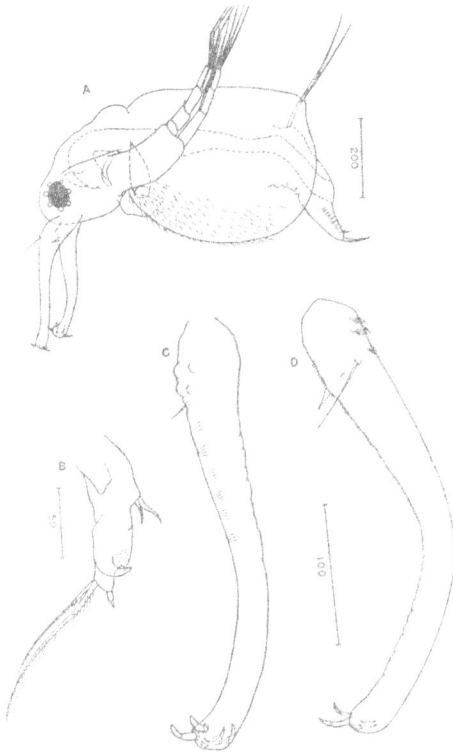

FIG. 16. *Moina affinis* Birge. *A.* Male from roadside ditch five miles east of Strong City, Kansas, on Old Highway 50, 18-VII-65. *B.* Male first leg from slide number 197*a* of Birge Collection; no locality or date given. *C.* Male antennule (medial view) from New Orleans, Louisiana, 18-IV-04 (Birge Collection slide No. M/14). *D.* Lateral view of same antennule as *C.*

measure from .7 to .8 mm. while males of *affinis* are seldom more than .6 mm. long.

Moina affinis is closely related to both *weismanni* and *flexuosa.* However, the three species are very restricted in their distribution, and it is highly doubtful that any of these three species would be found in the same region as the other. Although *weismanni* females have hairs on both the shell and head, these hairs are very sparse. *Moina flexuosa* females completely lack hairs.

DISTRIBUTION

Moina affinis was originally reported from Wisconsin (precise locality not given) by Birge (1893). It has subsequently been collected in all parts of the Central United States (fig. 20). The species has been reported as far east as Indiana but should be found along the East Coast of the United States. The Birge Collection included specimens of *affinis* from Bertig, Arkansas, and New Orleans, Louisiana. Specimens were also found in S. F. Light's collections from California.

In addition, the species has been found in rice fields in Italy by Moroni (1962) and Moroni and Vicini (1962). This is the only report of the species from the Old World so that it must be presumed that it is greatly restricted in its distribution and may have been recently introduced.

Specimens Examined

North America:
  Arkansas:
    Bertig, no date or precise locality given (Birge Collection)
  California:
    Elsinore, Riverside Co., three miles S. of Temecula, May, 1935 (S. F. Light Collection No. 78)
    Western border of Calaveras Co., large pond on Valley Springs, Stockton Road, 20 miles E. Stockton, 14-V-51 (S. F. Light Collection No. 328)
  Indiana:
    Roadside pond south of Bloomington, Monroe Co., 24-IX-60, collected by C. E. Goulden
  Kansas:
    Roadside pond east of Cheyenne Bottoms, Barton Co., 26-VII-58, collected by C. E. Goulden
  Louisiana:
    New Orleans, 6-IV-04 and 18-IV-04 (Birge Collections, slides 14–16, 29)
  Nevada:
    Small pond one mile N. Winnemucca, Humboldt Co. 7-IV-35 (S. F. Light Collection No. 193)
Europe:
  Italy:
    Cascina Corte Nuova rice fields, Parma, June 9, 16, 24, and 27, 1961 (collected by A. Moroni)

MOINA WIERZEJSKII RICHARD, 1895

Moina brachiata var. nov. Wierzejski, 1892: pp. 234–235; pl. 5, figs. 2–7.
Moina wierzejskii Richard, 1895: pp. 195–199; figs. 9–13.
Moina platensis Birabén, 1917: pp. 264–266; figs. 4–7.
Moina platensis Birabén, 1919: pp. 118–123; figs. 38–44.
Moina platensis Olivier, 1962: p. 216; pl. 13; figs. 2–5.
Moina wierzejskii Olivier, 1962: p. 217; pl. 14, figs. 1–3.

TAXONOMIC NOTES

This taxon is so variable in form that I thought at first it composed two quite distinct species. *Moina wierzejskii* was first described and illustrated by Wierzejski (1892) from Argentina as a variety of *brachiata* although he did not give it a new name. His figures and description clearly indicate that the specimens lacked the supraocular depression on the head, possessed a rather large pecten on the postabdominal claws, and had a two-egg ephippium. The male antennules had two sensory setae that originated very near the head, four terminal hooks; and the posterior margin of the antennules had a vertical row of long hairs.

As described by Wierzejski, the species is very distinct. In 1895 Richard described a species from Haiti that he thought to be similar to Wierzejski's variety. Since Richard recognized the form to be a new species, he proposed the name *Moina wierzejskii*. Richard's description, however, was not identical to that of Wierzejski's. For example, he made no mention of the very large claw pecten but instead illustrated a pecten of moderate size (fig. 17, *C*). He did not describe or illustrate the long hairs on the male antennule that are so apparent on Wierzejski's illustration.

I was able to examine plankton collections from the vicinity of Port au Prince, Haiti, that are part of Richard's material present in the Birge Collection and that may represent Richard's type material (he gave no specific type locality). This material was collected from 1885 to 1895. The specimens examined were identical to the form described by Wierzejski in that they possessed a large claw pecten and had the long hairs present on the posterior margin of the male antennule.

Simultaneous to the examination of Richard's Collections, I examined specimens of *Moina* from Kansas that were identical in all aspects to Richard's written description and illustrations of *M. wierzejskii*. These specimens possessed a much smaller claw pecten than the specimens from Haiti, and most of the males lacked long hairs on the posterior margin of the antennules. Because of the distinctly different claw pecten and the variation in pattern of hairs on the

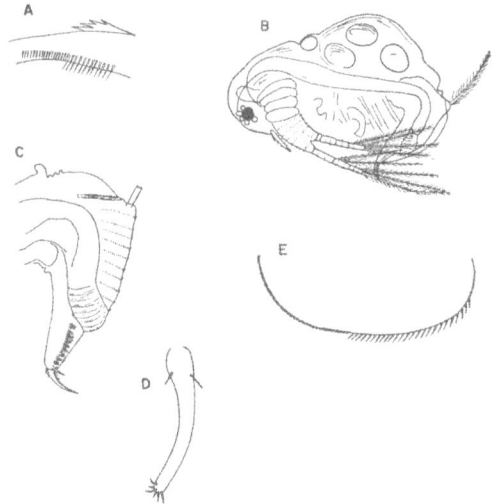

FIG. 17. Richard's original illustrations of *Moina wierzejskii* (after Richard, 1895). *A.* Postabdominal claw. *B.* Parthenogenetic female. *C.* Postabdomen. *D.* Male antennule *E.* Shell margin.

male antennules, it appeared that this form might be a distinct species.

However, after a long and careful study of these collections, I found a few specimens in Richard's material that differed from the majority of specimens in that they possessed claws with a small pecten. In some instances I found specimens in which one claw had a very large pecten while the second claw had a very small pecten. Likewise, similar specimens were found in the collections from Kansas (fig. 18, *E*).

It now seems apparent, after continued comparison of specimens from North America and Haiti, that there are two major forms of the species. One form, common in South America, Haiti, and the southern United States, normally has a large claw pecten. The second form, represented by the Kansas specimens and part of Richard's original material described in 1895, has a small claw pecten. Subsequently, I have found a third form of the species that will be described in the text.

DIAGNOSIS

Head and body either partially or completely covered with hairs. Head broadly rounded with a slight supraocular depression. The large eye lies near, but not contiguous to, the anterior and ventral head margins. The antennules are covered with hairs and rings of setae.

The ventral shell rim has twenty to twenty-six long setae that are followed behind by a row of short, ungrouped setae.

The postabdomen is ornamented behind with wavy rows of short setae and with rows of long hairs. There are nine to twelve lateral feathered teeth and one long bident tooth on the distal part of the postabdomen. The claw may have a very large pecten of seven to twelve long, stout teeth, or a small pecten of fifteen to twenty short, thin teeth.

The ephippium is ornamented with many protuberant round cells.

The antennules of the male are very stout, and the margin of the head is expanded laterally around the origin point of the antennules. The antennules have a faint bend very near the head and have the anterior margin lined with many short hairs. The posterior margin of the antennule has a vertical row of long hairs. The tip of the antennule has three or four short curved hooks and one long hook that points forward rather than medially as the other hooks do.

The first leg of the male has a greatly reduced hook and lacks the exopod with its long seta.

The postabdominal claw of the male has a pecten of many thin short teeth. The spermatozoa are small round cells.

DESCRIPTION

Richard (1895: pp. 195–198) described *Moina wierzejskii* as follows (fig. 17):

La longueur moyenne des femelles adultes [fig. 17, *B*] oscille autour de 1mm (en général elle est de 1mm15). La forme générale est lourde comme celle de *M. macropocus* Robin (= *M. paradoxa*). La tête, en particulier, est très analogue à celle de cette dernière espèce. Sa longueur est d'environ un tiers de la longueur totale. Il n'y a pas de sinuosité distincte au-dessus de l'œil, le bord dorsal de la tête présentant une courbe régulière. Le bord ventral de la tête est légèrement convexe. La tête est large par rapport aux valves, quand on examine des femelles dont la cavité incubatrice n'est pas distendue par de nombreux embryons. L'œil, de taille médiocre, est muni de lentilles crystallines globuleuses, grosses, assez nombreuses et assez bien dégagées du pigment. La fornix est très développée et s'étend jusqu'au-dessus de l'œil.

Les valves ont un bord ventral peu convexe. La convexité du bord dorsal est très variable suivant que les femelles ont peu ou beaucoup d'embryons. La moitié antérieure [fig. 17, *E*] de l'étendue totale du bord libre des valves (sauf la portion initiale du bord antérieur) porte des soies spiniformes courtes, espacées, peu nombreuses (on en compte de 20 à 25); toute la partie postérieure du bord libre est ornée de cils courts, serrés, devenant de plus en plus fins vers l'extrémité postérieure. Je n'ai pas pu découvrir de réticulation. Vers sa réunion à l'extrémité postérieure avec le bord opposé le bord postérieur de chaque valve présente un crochet opposé à celui de l'autre valve et entre les deux le bord dorsal de la carapace finit sur un petit limbe arrondi garni d'épines minuscules.

Les antennes antérieures subcylindriques, légèrement renflées à leur partie moyenne, ont environ la moitié de la longueur de la tête (mesurée de la naissance de la fornix à l'extrémité antérieure de la tête). Leur bord postérieur porte des cils grêles assez longs, très fins. Le bord anté-

rieur porte une soie fine plus courte que la moitié de la longueur de l'antenne. Sur la surface des antennes on peut distinguer, à un très fort grossissement, de très fines spinules disposées par séries transversales. Il y a à l'extrémité un bouquet de 8 à 9 soies sensorielles très courtes.

Les antennes de al deuxième paire n'ont rien de particulier. Elles sont robustes et les articles sont garnis de très fines spinules.

Les pattes de la première paire sont constituées comme celles de *M. rectirostris* et ont 12 soies disposées comme dans cette espèce. Mais les deux soies insérées l'une à l'extrèmité supérieure du premier article et l'autre à l'extrémité supérieure du deuxième ne sont pas semblables aux autres. Elles paraissent uniarticulées et ont la forme de longues épines très finement et brièvement barbelées. Les autres soies sont longuement et finement ciliées, la plus rapprochée de la deuxième épine est plus longue que les autres. Les deux soies postérieures de la face antérieure sont recourbées à leur extrémité et garnies de cils courts.

Le postabdomen [fig. 17, *C*] a la forme ordinaire. La longueur de la partie conique forme le tiers de la longueur du postabdomen (entre la naissance des griffes et l'origine des soies postabdominales). Il porte 9 ou 10 dents simples, aiguës, barbelées et une dent bifide longue, étroite, dont la branche proximale est très courte. Les griffes terminales bien développées et assez incurvées [Fig. 17, *A*] portent à leur base un peigne de 12 à 15 dents bien distinctes. Le reste est nettement et densément cilié jusqu'á l'extrémité. A l'origine de la face ventrale de chaque griffe est une expansion chitineuse divisée en dents et qui se montrent sur la griffe vue de profil comme des dents dirigées en arrière. Les voies postabdominales sont longues.

Beaucoup des femelles examinées étaient éphippiales. L'éphippium a deux loges et est réticulé sur tout son étendue.

Le *mâle* mesure de 0mm75 à 0mm80. Il se distingue immédiatement des mâles connus jusqu'ici par la structure de ses antennes antérieures.

Le bord dorsal des valves est droit. Les soies de leur bord ventral sont disposées comme chez la femelle. Le nombre des dents du postabdomen est le même (8 ou 9). Le peigne des griffes est formé de dents plus fines, plus nombreuses que chez la femelle (environ 20 à 30) et resemble plutôt à des cils. Les dents de la ligne chitineuse de la naissance (face ventrale) des griffes sont aussi plus nombreuses. Le reste de la griffe paraît lisse.

Les antennes antérieures [fig. 17, *D*] sont insérées près du sommet de la tête, à la hauteur de l'œil; elles sont légèrement renflées à la base, et régulièrement incurvées, à concavité interne, sans présenter la géniculation bien distincte et à angle net des autres espèces. C'est à l'extrémité du premier cinquième de sa longueur que l'antenne présente une courte épine à la face antéro-interne et une soie grêle à la face antéro-externe. L'extrémité à peine renflée de l'antenne porte, outre un petit nombre de soies sensorielles très courtes et semblables à celles de la femelle, quatre (quelquefois cinq) crochets chitineux plus ou moins recourbés, généralement peu aigues. L'un d'eux est presque toujours plus long que les autres et peut atteindre 0mm05. Les pattes de la première paire sont exactement comme les représente Wierzejski. Des deux soies en aiguillon de la femelle, l'une est ici un fort crochet recourbé, l'autre forme un crochet lisse, pâle, conique, incurvé de façon à ressembler à une griffe terminale du postabdomen. La soie ciliée qui, chez la femelle, est plus longue que les autres, ne dépasse guère la taille ordinaire chez le mâle, et sa ciliation est aussi la même que dans les autres soies. (Richard, 1895: pp. 195–199.)

Female

The head is broadly rounded, with an indication of a slight supraocular depression which is not pronounced in most forms (fig. 18, *A*). The body is rotund and the brood sac may be so expanded as distinctly to separate the head from the shell.

The eye is large and lies near the antero-ventral margins of the head but is not contiguous with them. The antennules are large and robust in appearance. The sensory seta orignates from a point just distal to the middle of the antennule (fig. 18, *C*). There are several rings of short setae that extend from the base of the antennules to the tip. There is also a vertical row of groups of long hairs, three or four hairs in each group. The tip of the antennules has nine to ten sensory papillae.

The pattern of hairs on the head and shell is variable. Specimens from Haiti, Texas, and central California have hairs covering the posterior half of the head and the anterior part of the shell. Specimens from Arizona and Oregon have hairs completely covering the head and shell while specimens from Kansas have hairs only on the posterior half of the head but completely covering the shell.

The shell is heavily reticulated and has a row of twenty to twenty-six setae along the ventral shell rim that are followed behind by a row of much shorter setae not arranged in groups. The dorsal-posterior corner of the valves carries a pair of hooks that are not rounded but instead are elbow-shaped hooks as in *macrocopa*.

The setation pattern of the first leg is like that of most species of *Moina*; the anterior seta of the penultimate segment is very stout and lacks hairs or teeth (fig. 19, *D*). The anterior seta of the terminal segment is also very stout but does have an inner row of fine short hairs near its terminal end. The other two terminal setae are feathered.

The postabdomen is a very robust structure that is equal in length to about one-third the total body length (fig. 18, *B* and *D*). The dorsal region of the postabdomen is ornamented with wavy rows of short setae that interconnect to form a network. There are horizontal rows of long hairs that extend from one side to the other side of the postabdomen and are quite noticeable along the dorsal margin. In some specimens, these hairs may be very numerous and are not arranged in rows; this is true of specimens from central California and Arizona.

The distal, conical part of the postabdomen has from nine to twelve lateral feathered teeth and one long bident tooth. The claw is armed with a pecten of seven to twenty teeth (fig. 18, *D–E*). The teeth of the pecten may vary in length from rather short teeth that just extend beyond the claw base, to long stout teeth that are equal in length to the width of the claw. As stated above, this variation initially seemed to indicate that there were two distinct species, i.e., the typical *wierzejskii* form from Haiti, South America, and the Southern United States which had a large pecten of only seven to twelve teeth and the form from Kansas that had a rather short pecten of about twenty teeth. I was later able to determine that the size and number of teeth of the pecten depended on the age of the female; immature specimens may lack this pecten completely or have only a short pecten. However, the pecten has become permanently small in specimens from Kansas which suggests that a smaller pecten has some adaptive significance in Kansas forms, or else that this is the primitive form in the species. I have found specimens in various collections that have both a small and a large pecten on the two claws of the same postabdomen (fig. 18, *E*).

The ephippium has two sexual eggs and is ornamented with a cobble stone pattern of protuberant cells (fig. 18, *A*). These cells appear to be slightly separated in this species, not close together as in *australiensis* and *brachycephala*.

The females examined by Wierzejski (1892) measured up to 2.67 mm. long while Richard's specimens were 1.15 mm. long. The parthenogenetic females that I have measured range from 1.08 to 1.47 mm. and the sexual females from 1.05 to 1.17 mm. Birabén's specimens of *platensis* from Argentina were 1.24 to 1.71 mm. long.

It seems improbable that specimens can be as large as 2.67 mm. long, as reported by Wierzejski (1892); instead of having a maximum length of 1.7 or even 1.8 mm which would seem more likely.

Male

The males of *wierzejskii* are very characteristic and should not be confused with those of any other species (fig. 19, *A*). The head is rather broad and is flattened in front. The antennules originate from the antero-ventral corner of the head, and the lateral contour of the head bulges on either side of the antennules. The eye is large and almost fills the tip of the head.

The head and shell may or may not be covered with hairs; this depends on the geographical form of the species. The males from Haiti have hairs only on the shell; whereas males from California, Texas, and Kansas are almost completely covered with hairs.

The antennules have a slight bend near the proximal end, and the two sensory setae originate from this point. The short seta is located on the medial margin; whereas the long thin seta originates on the lateral side behind the other seta. The tip of the antennules has four or five hooks, one of which may be longer than the others and points forward rather than medially as the other hooks do. There is usually a row of short but densely packed hairs all along the anterior margin of the antennules. The posterior margin, in contrast, has a row of long hairs that, however, are not densely

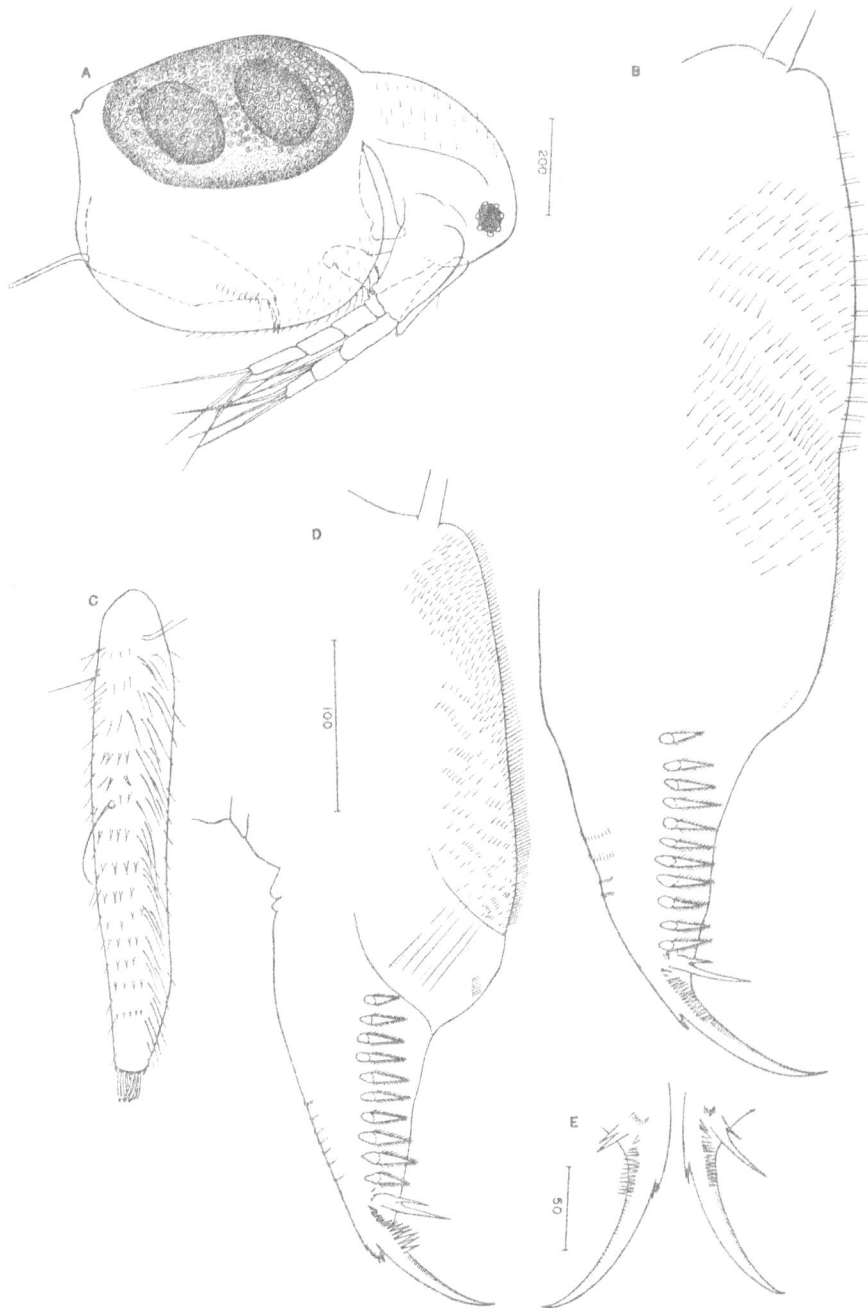

Fig. 18. *Moina wierzejskii* Richard. *A*. Sexual female from roadside ditch along county road one mile east of Cheyenne Bottoms Waterfowl Refuge, Barton County, Kansas, 26-VII-58. *B*. Postabdomen of *A*. *C*. Antennule of *A*. *D*. Postabdomen of female from pool between Globe and Miami, Arizona, May, 1935 (S. F. Light Collection Number 133 (1)). *E*. Right and left postabdominal claw of same female from roadside ditch along county road one mile east of Cheyenne Bottoms Waterfowl Refuge, Barton County, Kansas, 26-VII-58.

packed and may be found only near the distal end of the antennules or may extend from the proximal to the distal end. Again, this varies among the geographical variants of the species.

The second antennae are behind both the eye and the antennules. They differ little from the female antennae, although the two rami are slightly shorter.

The shell may either be completely covered with hairs or may only have hairs near the ventral and anterior shell rim. As in the female, the shell is reticulated and the ventral rim has a row of twenty to twenty-five setae. There is a row of very short, un-arranged hairs or setae behind the long setae on the ventral rim.

The first leg of the male is considerably reduced in size—smaller than any other species among the large species of *Moina* (fig. 19, *C*). The hook is also reduced in size but the penultimate segment, from which it originates, is elongated and has very few hairs on its medial margin. However, the terminal segment bears a hooklike seta that in all other species may only be a

thin curved seta but in *wierzejskii* is a strong hook that is larger than the hook on the third segment. There is no exopod on this foot. This is a unique character that *wierzejskii* shares with *hutchinsoni*, the only two species that produce two sexual eggs yet lack the exopod on the first leg of the male.

The postabdomen of the male is like the female postabdomen. The vas deferens opens on the side of the postabdomen ventral to the second or third proximal feathered tooth (fig. 19, *B*). The claw has only a small pecten and in some instances no pecten. The pecten, when present, is never as large as that of the female—this is true of all geographical races of the species whether the female has a large or small pecten.

The spermatozoa are small round cells.

Moina wierzejskii demonstrates considerable infra-specific variability, but there is little reason to believe that its forms should be divided into distinct species. The general body morphology is consistent. The head and body may or may not be covered with hairs, specimens from Haiti, Texas, and Kansas have hairs only on part of the head; whereas specimens from Arizona and California have the head completely covered with hairs. The teeth of the claw pecten differ considerably in size and number. Adult specimens from Haiti, Texas, Arizona, and California have a pecten of seven to twelve long, stout teeth, but specimens from Kansas and all immature individuals of the species have a claw pecten of about twenty short thin teeth. Some Kansas specimens may be seen to have both the large pecten and the smaller one on the two claws of the same individual.

The males may differ according to the density of hairs on the antennule. Some specimens have a row of long hairs covering the entire posterior margin of the antennule while others have these hairs only near the tip—but this character seems to differ with age of the males. However, most specimens have only four terminal hooks on the antennules. Some specimens from Kansas have five hooks, and Richard states that there were five hooks on some male specimens from Haiti.

It would seem that the variation in the species is due, in a large part, to adapation to the local environment. Perhaps these forms deserve subspecific rank; however, I am reluctant to apply names to each group until additional study of the amount of variability within each group. Collections of this species have been placed in the United States National Museum. Part of Richard's material from Haiti has been deposited in the British Museum (Natural History).

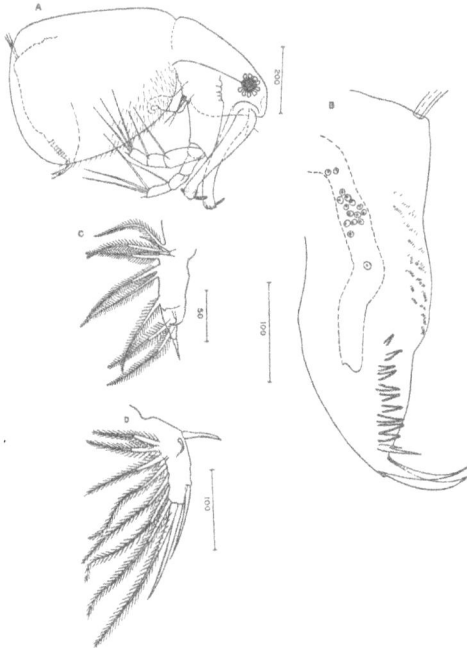

FIG. 19. *Moina wierzejskii* Richard. *A*. Male from Mare T. Vienp. et Briqueterie, Port au Prince, Haiti, 1-V-1896 (Birge Collection No. 800). *B*. Postabdomen of male from roadside ditch five miles east of Highway 281 at Junction of N-5 County Road and Salt Marsh Road, Stafford County, Kansas, 20-VI-65. *C*. First leg of *A*. *D*. Female first leg, roadside ditch along county road one mile east of Cheyenne Bottoms Waterfowl Refuge, Barton County, Kansas, 26-VII-58.

DIFFERENTIAL DIAGNOSIS

Moina wierzejskii is restricted in its distribution to the New World; it occurs only within the geographical ranges of *macrocopa* and *brachycephala* among the large species of *Moina* and of *affinis* and *micrura*

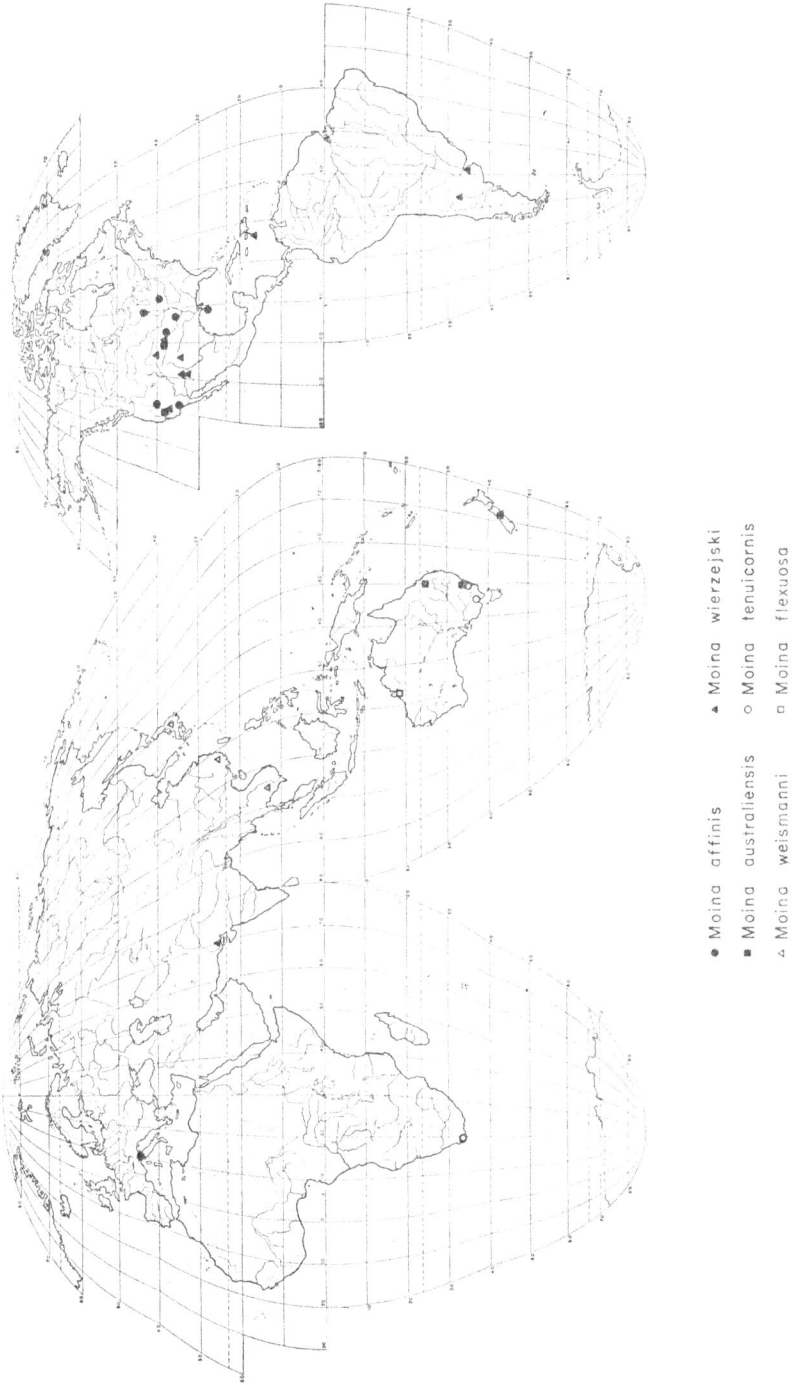

FIG. 20. Geographical distribution of *Moina affinis, wierzejskii, australiensis, tenuicornis, weismanni,* and *flexuosa.*

• Moina affinis　　　　▲ Moina wierzejski

■ Moina australiensis　　○ Moina tenuicornis

▲ Moina weismanni　　　□ Moina flexuosa

among the small species. I have collected *macrocopa* and *wierzejskii* together from roadside ditches in Kansas. It should not be too difficult to distinguish these two species, however, since females of *macrocopa* have the unique first leg while the males have the antennules bent at the mid-point rather than at the base as in *wierzejskii*. The ephippia also are different —in *M. macrocopa* the ephippium is reticulated with rectangular cells while the ephippium of *M. wierzejskii* has many raised knoblike protuberant cells covering the surface.

Moina brachycephala has a very large head and has the eye very near the center of the head. In *M. wierzejskii* the head is not so large, and the eye is almost contiguous to the margin. The ephippium of *M. brachycephala* is also ornamented with the knoblike protuberances, but the individual cells are closer together than in *M. wierzejskii*. The males of *brachycephala* have unique antennules; the sensory setae originate at a point about one-fourth the distance from the head, and there are three accessory hooks on the distal end of the antennules. The first leg of the male of *brachycephala* has an exopod which is absent in *wierzejskii*.

Moina micrura and *affinis* are smaller than *wierzejskii* and both have a well-developed supraocular depression and a narrow head. *Moina micrura* lacks hairs on the body while the male has the antennules bent one-third the distance from the head and has a large hook on the first leg.

Moina affinis is quite similar to *wierzejskii* in many respects. The males, in fact, have the antennules and first leg so much alike that it tends to suggest a close relationship between the two species. However, the males of *affinis* are much smaller than the males of *wierzejskii*.

Moina hutchinsoni also shares many features with *wierzejskii* but can be distinguished by the absence of the bident tooth on the postabdomen. The former species is restricted to the very saline and alkaline lakes of Western North America.

DISTRIBUTION

Moina wierzejskii has been reported from North and South America as well as from Haiti in the West Indies (fig. 20). In the United States it is known from Arizona, California, Colorado, Kansas, and Texas. In South America, it is known only from Argentina (La Plata, according to Wierzejski, 1892—Mendoza, according to Birabén, 1917, 1919).

The species is therefore restricted to the New World and, with the exception of Haiti, is found only in temperate zones of North and South America. Furthermore, it seems to be almost totally restricted to the arid and semi-arid regions where it is found in great abundance in small temporary pools. It is possible that the species will eventually be found in

Central America and in other parts of South America, but this does not seem likely because of its affinity for dry regions.

Specimens Examined

North America:

Arizona:

Between Globe and Miami, 75 miles from Phoenix, Gila Co., small ponds by side of road, May 1935 (S. F. Light Collection No. 133(1))

Red Lake, 24-VII-33, collected by G. E. Pickford

Colorado:

Hugo, 31-V-05 (Birge Collection No. 24 and 26)

California:

Merced Co., 10 miles S. of Merced, roadside pond, May, 1935 (S. F. Light Collection No. 51)

Kansas:

Roadside pond, Barton County, 10 miles N. E. Great Bend, 26-VII-58, collected by C. E. Goulden

Texas:

Lubbock, Mud from Playas near town, collected by Vernon Proctor, 1-XI-64

Haiti:

Mare Coucher, Port au Prince, 1-V-1896 (Birge Collection No. 751, from Richard's collections)

Mare T. Vienp. et Briqueterie, Port au Prince, 1-V-1885 (Birge Collection No. 800, from Richard's collections)

Mare Brinville, Port au Prince, April, 1895 (Birge Collection No. 752, from Richard's collections)

MOINA AUSTRALIENSIS SARS, 1896

Moina australiensis Sars, 1896: pp. 18–24; pl. 3, figs. 1–11.

Moina australiensis Henry, 1922: pp. 34–35.

Moina australiensis Brehm, 1953b: pp. 6–9; figs. 2a–e.

DIAGNOSIS

The parthenogenetic females are 1.25 to 1.55 mm. long, the males .08 mm. long. The head and anterior part of the shell are covered with thin hairs, and the head has a supraocular depression. The antennules are long and thin and have several rows of setae on them. The shell has a row of approximately thirty long setae on the ventral margin. Behind these setae, and along the posterior margin, the shell rim is ornamented with groups of shorter setae, each group composed of five to eight unequally sized setae.

The postabdomen has ten to twelve lateral feathered teeth and one long bident tooth. The claw has a pecten of about twenty short thin teeth.

The ephippium has two sexual eggs. It is ornamented with a cobblestone pattern of round, knoblike cells.

The male antennules are bent near the head and have four terminal hooks. The male first leg has a well-developed hook on the penultimate segment and a moderately long seta on the exopod.

DESCRIPTION

Sars' original description of this species is the only extensive description that has been published; and in fact, there are few published records of *australiensis*. As a result, the species is poorly known. These are Sars' comments (fig. 21):

The largest specimens examined attain a length of 1.30 mm., and this form accordingly grows to a somewhat larger size than *M. propinqua*, which, as a rule, does not exceed a length of 1 mm.
The general form of the body [fig. 21, *B*] is that characteristic of the genus, the head being very sharply marked off from the shell by a deep dorsal depression.

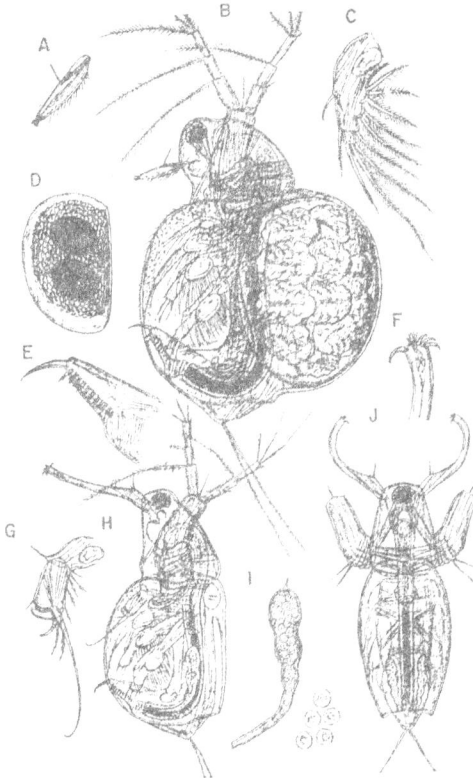

FIG. 21. Sars' illustrations of *Moina australiensis* (from Sars, 1896). *A*. Antennule. *B*. Parthenogenetic female. *C*. Female first leg. *D*. Ephippium. *E*. Postabdomen. *F*. Tip of male antennule. *G*. Male first leg. *H*. Male. *I*. Testis and spermatozoa. *J*. Dorsal view of male.

The form of the shell appears rather variable, according to the degree of distention of its dorsal part with ova or embryos. Sometimes this part is quite enormously distended, so as to form an almost globular expansion, sharply defined from the valvular part of the shell, and this is generally the case with all the individuals of the earlier generations. The valvular part of the shell, however, preserves its shape unaltered, in all specimens being comparatively small, so as not fully to obtect the tail, a greater part of which is always seen to project freely beyond the shell posteriorly. At the junction between the dorsal and valvular parts, the shell projects posteriorly as a short and obtuse prominence, below which the posterior edges appear slightly incurved. The inferior edges of the valves are nearly straight, and join by a perfectly even curve both the anterior and posterior edges. They are clothed with small marginal hairs, which are more distinct in the anterior part, gradually disappearing behind.

The head is comparatively small and somewhat procumbent without any trace of a dorsal crest, being evenly vaulted above. Just above the ocular region, a slight sinus may be traced, but this sinus is not nearly so pronounced as is *M. propinqua*. The frontal, or ocular part is but slightly prominent, and is evenly rounded; and the ventral edge of the head forms only a slight convexity at the insertion of the antennulae, without being defined by any perceptible notch from the base of the labrum. Of the fornix, only a slight trace is found as a somewhat elevated ridge above the base of the antennae.

The eye is of moderate size, and provided with a number of well-defined crystalline bodies. The ocellus, as in the other species of the genus, is wholly absent.

The antennulae [fig. 21, *A*] are comparatively short, scarcely exceeding half the length of the head, and exhibit a somewhat fusiform shape, being distinctly dilated in the middle. They are, as in the other species, freely mobile, and are clothed posteriorly with delicate cilia. Anteriorly each antennula carries a single sensory bristle, occurring somewhat nearer to the base than to the tip, and the latter has a bundle of very small olfactory papillae.

The antennae are powerfully developed, and of the structure characteristic of the genus. The scape is very massive and strongly muscular, and is provided near the base outside with 2 juxtaposed and rather long sensory bristles, which, especially in the dorsal or ventral views of the animal, are very conspicuous; at the end, another sensory bristle is seen projecting between the bases of the rami. Both the outer part of the scape and the rami are densely hairy, and the natatory setae finely ciliated.

The 1st pair of legs [fig. 21, *C*] are constructed in the very same manner as in the European species, *M. brachiata*, and differ markedly from those in *M. paradoxa*, by the sub-apical seta being quite simple and finely ciliated, like most of the other setae, whereas in *M. paradoxa*, according to the statement of Prof. Weismann, this seta is very strong, spiniform, and coarsely denticulated anteriorly. The tail agrees in its structure with that in most other species, its outer part beyond the anal opening [fig. 21, *E*] being conically tapered, and provided on each side with 10–12 denticles, the outermost of which is bidentate, whereas the others are extremely delicate, squamiform and finely ciliated on both edges. The terminal claws are perfectly smooth, without any trace of secondary teeth.

The ephippium [fig. 21, *D*] is of an oval or somewhat semilunar form, and has the surface very coarsely reticulated. It contains, as in *M. paradoxa*, 2 egg-ampullae, which are somewhat obliquely disposed, the one behind the other. Before the ephippium is detached from the shell, however, the 2 winter-eggs occupy a rather different place, being, as shown in the succeeding specie, juxtaposed in the anterior part of the matrix; and in a lateral view of the animal it therefore appears as if only a single ovum were present.

The adult male [fig. 21, H, J] is rather inferior in size to the female, scarcely exceeding a length of 0,80 mm., and exhibiting a very different appearance.

The shell is much narrower, its dorsal part not being at all expanded; and the posterior extremity appears, in the lateral view of the animal, obtusely truncated, forming above almost a right angle. The inferior edges of the valves are densely clothed with fine hairs, which, in their anterior part, assume a fur-like appearance.

The head looks very different from that in the female being much longer and more erect. It gradually tapers towards the front, which appears obtusely truncated and defined above by a very slight sinus. The inferior edge of the head is perfectly straight and horizontal, whereas the upper one is obliquely ascending and, but very slightly convex.

The antennulae, which issue from the most anterior part of the head, are greatly developed, being fully as long as the head. They are very mobile, but, as a rule, extended obliquely anteriorly, with the terminal part more or less incurved [fig. 21, G]. Near the base they each form a somewhat genicular bend, and at this place 2 unequal sensory bristles are seen to project anteriorly. The outer part of the antennulae is rather narrow, and nearly cylindric, and terminates with 4 strongly curved hooks, between which a small bundle of olfactory papillae is traceable [fig. 21, F].

The 1st pair of legs [fig. 21, G] are, as usual, transformed into powerful grasping organs. They closely resemble in structure those in the male of *M. paradoxa*, as figured by Prof. Weismann, each having a rather long, setiform appendage extending beyond the claw, and terminating in a fine hooked point. In addition, a thin lamella is seen projecting inside the claw, having 3 apical bristles, the anterior of which is curved in a hook-like manner, and devoid of cilia. (Sars, 1896: pp. 18–22.)

Sars apparently did not recognize the hairs that are present on the head and anterior part of the shell in his specimens although in describing the male he states that the ventral edges of the valves are densely clothed with fine hairs "which, in their anterior part, assume a fur-like appearance." This reference is not to the marginal setae, but Sars was actually observing the hairs on the shell. In the male, these hairs are so densely packed near the ventral shell rim that they obscure the marginal setae.

Female

The head has a supraocular depression almost as deep as the one that may be seen on heads of *Moina brachiata*. The antennules, rather than "comparatively short" as stated by Sars, are instead rather long in comparison to other species of the genus. They are almost as long as the basipod of the second antenna.

The ventral shell rim has a row of about thirty setae which are followed by several groups of smaller, but unequally sized setae on the posterior rim. The setae on the posterior shell rim are in groups, each group consisting of five to eight setae that increase in size posteriorly. In addition, there is a pair of elbowlike hooks at the dorsal end of the posterior shell margin.

The dorsal margin of the postabdomen is covered with rows of short setae and has a few long hairs near the margin. The distal part of the postabdomen has ten to twelve long, feathered teeth. In addition, the bident tooth is very long.

The postabdominal claw has one or two thin teeth or "Basaldorn" on the ventral margin and a distinct pecten of approximately twenty teeth on the opposite side. This pecten is composed of thin teeth that barely extend beyond the dorsal margin of the claw. The distal part of the claw margin has a row of short setae.

The ephippium of the sexual female contains two sexual eggs. It is ornamented with small round cells producing a cobblestone pattern.

Male

The male, like the female, is covered with long hairs. These hairs are absent from the posterior part of the shell. The antennules are very long and have the two sensory setae located near the head. The distal end of the antennule has four hooks.

The first leg of the male has a moderately developed hook. The hook does not fold back onto the basal segment of the leg but curves ventrally and medially. The terminal seta on the exopod segment is not as long as that of *macrocopa* or *belli* and does not extend to the posterior margin of the shell.

The male postabdomen is like the female's. The vas deferens opens on the side of the postabdomen near the ventral margin—not near the anal opening as described by Sars (1896: p. 22). The spermatozoa are small round cells. The type material of this species is in the Zoologisk Museum, Oslo, Norway. Additional material from Australia has been placed in the United States National Museum.

DIFFERENTIAL DIAGNOSIS

There are few species with which *australiensis* may be confused, not only because of its characteristic body morphology but also because it has a very restricted distribution (Australia and New Zealand; fig. 20). Only *Moina tenuicornis* and *micrura* are found in the same region. *Moina tenuicornis* can be distinguished by the long thin antennules, by the absence of a supraocular depression, the ungrouped setae on the posterior shell rim and the large pecten on the claw. *Moina australiensis* has a supraocular depression, groups of setae on the posterior shell rim, and has a very small pecten on the postabdominal claw. The males of these two species can be distinguished only with difficulty, however, and one must rely on comparison of the relative length of the antennules for separation. Otherwise, the basic body form and structure of the first leg and postabdomen are identical.

Moina micrura is a smaller species than *australiensis*. In addition, the postabdomen of *australiensis* has long hairs along the dorsal margin as well as on the head and shell. These hairs are completely lacking in *micrura*. The males of these two species are very different.

The female of *M. australiensis* is strikingly similar to the female of *Moina brachiata* in body form and size. However, as *micrura*, *brachiata* females lack long hairs on the dorsal surface of the head while *australiensis* females, on the other hand, lack the very large claw pecten of *brachiata*.

DISTRIBUTION

According to Henry (1922), *Moina australiensis* has been reported from "Sydney, Kensington, the Waterloo Swamps and ponds near Bourke St. and Botany Road. It also occurs in Victoria." Brehm (1953b) found this species in a brackish water pond near Tammin in East Australia. Sars' slide material contains a few specimens from New Zealand (precise locality not given). In addition, I have material collected by Dr. Vida Stout from ponds near Christchurch, New Zealand.

Specimens examined

Australia:

Sydney, New South Wales (Sars Collection No. 53.2/12)

Waterloo Swamps near Sydney, New South Wales, collected by Ramsey, 1888 (Sars Collection)

New Zealand:

No locality or date given (Sars Collection—determined as *M. propinqua*)

Christchurch (Vida Stout).

MOINA TENUICORNIS SARS, 1896

Moina tenuicornis Sars, 1896: pp. 24–27; pl. 4, figs. 1–8.

Moina tenuicornis Sars, 1916: pp. 320–321; pl. 35, figs. 2, 2a–c.

Moina tenuicornis Henry, 1922: p. 35.

DIAGNOSIS

Head and shell completely covered with hairs. Head without a supraocular depression. Female with very long, thin antennules—longer than the basipod segment of the second antennae. Antennules originate from a distinct prominence on the ventral margin of the head. Ventral margin of shell with a row of about forty setae. Posterior shell margin ornamented with a row of equally sized but very short setae that are not arranged in groups. Postabdomen with ten to twelve lateral feathered teeth and one long bident tooth. Claw with a distinct pecten of approximately twenty long thin teeth.

Antennules of the male long and thin, bent near the ventral margin of the head, and with four hooks on the distal end. First leg of male has a long, well-developed hook and a moderately long seta on the exopod.

Ephippium with two eggs and ornamented with rectangular reticulations.

DESCRIPTION

Sars' (1896) original description and figures serve completely to characterize the species (fig. 22):

The length of the largest specimens is 1,20 mm., and this form is accordingly but little inferior in size to the preceding one [*Moina australiensis*].

The general form of the body [fig. 22, *A*] agrees rather closely with that in *M. australiensis*. On a closer comparison, however, the head is found to differ very markedly in shape, being quite evenly vaulted dorsally, without exhibiting any trace of a sinus above the eye. The front is obtusely rounded and, as in the preceding species, but little prominent. On the other hand, the inferior part of the head is produced, at the insertion of the antennulae, to a rather conspicuous, rounded prominence, which is defined behind by a deep notch, giving the head, in a lateral view of the animal, a physiognomy rather different from that in *M. australiensis*.

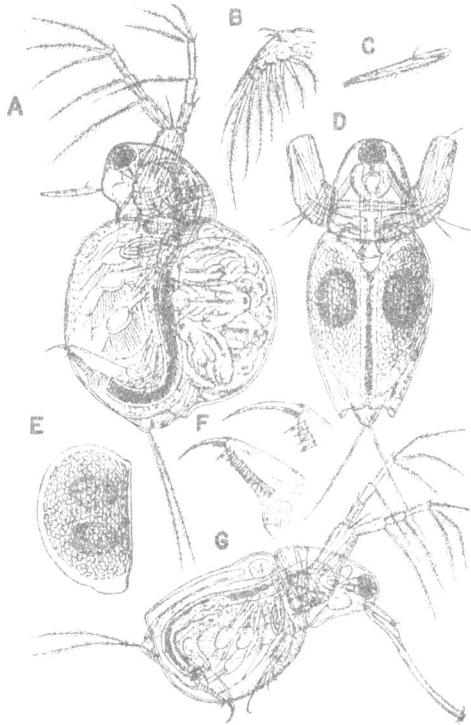

FIG. 22. Sars' illustrations of *Moina tenuicornis* (from Sars, 1896). *A.* Parthenogenetic female. *B.* Female first leg. *C.* Female antennule. *D.* Dorsal view of sexual female. *E.* Ephippium. *F.* Female postabdomen. *G.* Male.

In the form and structure of the shell, scarcely any essential difference is to be found from that in the said species, excepting that perhaps the inferior edges of the valves are more densely setiferous.

The eye is comparatively larger than in *M. australiensis*, almost completely filling up the frontal part of the head, and the crystalline bodies seem also to be more numerous.

The antennulae [fig. 22, *C*] are distinguished by their unusual length, being comparatively nearly twice as long as in the preceding species. They are very narrow, sublinear in form, and provided behind with scattered, delicate hairs. The sensory bristle of the anterior edge occurs much nearer to the base than to the tip, being placed at about the end of the first third part of the length of the antennula.

The antennae do not exhibit any essential difference from those in the preceding species; the 1st pair of legs [fig. 22, *B*] too are constructed in the very same manner, the subapical seta being, as in that species, quite simple.

The tail [fig. 22, *F*] likewise looks rather similar. But, on a closer examination, the terminal claws are found to differ essentially in being each provided at the base with a well-marked, comb-like series of secondary teeth [fig. 22, *F*].

A similar armature, as is well known, is found in *M. brachiata*, whereas in the other European species, *M. paradoxa*, the claws are quite smooth, as in *M. australiensis* and *M. propinqua*.

The ephippium [fig. 22, *E*], as in the preceding species, always contains 2 egg-ampullae, which in this form, however, are placed more transversely, or in a manner similar to that figured by Prof. Weismann in the ephippium of *M. paradoxa*. Before the ephippium is detached from the shell, the 2 winter-eggs occupy [fig. 22, *D*] a similar juxtaposed situation in the matrix to that observed in the preceding species.

The adult male [fig. 22, *G*] scarcely exceeds a length of 0.70 mm., and on the whole closely resembles the male of *M. australiensis* though the antennulae appear still more elongated, considerably exceeding half the length of the body. The 1st pair of legs are constructed in the very same manner as in the male of the said species, and the zoosperms in this form also are represented by simple, clear, nucleated cells. (Sars, 1896: pp. 24–27.)

Female

The head of the female is broadly rounded and is completely covered with hairs. The antennules are very long and thin; the width is approximately one-twentieth of the length. The antennules are longer than the basipod of the second antennae (see fig. 22). The sensory seta of the antennule originates laterally and at a point one-third the distance from the head. The antennules have rows of short, stout, setae encircling the distal two-thirds of the antennule, and there is a vertical row of long fine hairs extending from the base to the distal end.

The shell of the female is completely covered with hairs except on the ephippium of the sexual female. The shell surface is reticulated with a network of vertical lines connected by short horizontal lines.

The ventral shell rim has a row of approximately forty setae that extends almost to the posterior margin of the shell. There is a row of shorter setae along the

posterior margin. These short setae are not organized into groups.

The dorsal margin of the postabdomen is ornamented with rows of short setae and several long, thin hairs. The distal conical portion of the postabdomen has from ten to twelve lateral feathered teeth and one long bident tooth. The claw has a distinct pecten of about twenty teeth. This pecten is only slightly larger than that found on specimens of *Moina australiensis*.

The ephippium of the sexual female is reticulated with rows of rectangular cells. It contains two sexual eggs.

Male

The male is almost identical to the male of *australiensis* but has slightly longer antennules. The antennules have four distal hooks. The first leg of the male is identical to that of *australiensis*.

The spermatozoa are small simple nucleated cells not at all like those of either *macrocopa* or *brachiata*. The genital opening lies on the side of the postabdomen near the ventral margin.

The type material of *M. tenuicornis* is in the Zoologisk Museum, Oslo, Norway.

DIFFERENTIAL DIAGNOSIS

The major characters that differentiate *tenuicornis* from *australiensis* are the absence of a supraocular depression, the longer antennules in both male and female, the presence of hairs covering the entire body, the setae pattern on the posterior shell margin, and the longer claw pecten. The ephippium of *tenuicornis* has rectangular reticulations while that of *australiensis* has round knoblike cells.

Since these two species are so morphologically similar, it seems quite unusual that they should coexist in the same geographical region unless there are strong physiological or ecological differences. Unfortunately, nothing is known of their ecological requirements.

DISTRIBUTION

Type locality: "Pond at the corner of Bourke Street and Botany Road, Sydney, Australia" (Sars, 1896: p. 27).

This species is largely restricted to Eastern Australia (fig. 20). Sars has identified a form from the Union of South Africa that he thought to be *Moina tenuicornis*. I have examined one of his specimens that had been dissected and placed on a slide. Unfortunately, the slide had partially dried and was damaged. Therefore, I could not determine whether or not the specimen was identical to Sars' Australian specimens. However, his figures of the South African form are so similar to his original *Moina tenuicornis* that it is highly probable that it is the same form.

Specimens Examined

Australia:

Victoria: No date or locality given (Sars Collection No. 53.2/11)

Victoria: St. Arnauds (Sars Collection No. 53.2/12)

New South Wales: Sydney, no date given (Sars Collection No. 53.2/12) Black Rock, no date given (Sars Collection No. 53.1/5) Glenfield, 16-III-60; irrigation ditch (collected by N. J. Williams—sent to the author by Ian Bayly).

MOINA WEISMANNI ISHIKAWA, 1896

Moina weismanni Ishikawa, 1896: pp. 1–6; pl. 2, figs. 1–9.

Moina brevicornis Sars, 1903b; pp. 10–12; pl. 1, figs. 3, 3a–b.

Moina weismanni Ueno, 1927: p. 284; pl. 25, figs. 14, 14a–b; pl. 26, figs. 14c-f.

Moina weismanni Brehm, 1951: pp. 112–113; figs. 34–36.

Moina vom Mandvi-fleuve, Brehm, 1953a: p. 331; figs. 95a-b.

DIAGNOSIS

The females are up to 1.0 mm. long. The head has a supraocular depression. The antennules are short and spaced far apart on the flat ventral margin of the head. The shell is rounded and with about seventeen long setae on the ventral shell rim followed behind by short setae that are arranged in groups. The postabdomen has seven to nine feathered teeth plus a bident tooth that has both arms quite long. The claws have a pecten of fine teeth. The ephippium has one egg and is ornamented with large circular cells that are raised above the shell surface.

The sensory setae on the male's antennules are near the head. There are four terminal hooks on the antennules. The shell of the male is covered with hairs. The first leg has a small hook on the third segment and a long, hooklike seta on the terminal segment. The spermatozoa are small spherical cells.

DESCRIPTION

I have not seen Ishikawa's original material but have examined Sars' type material of *M. brevicornis*. Sars' specimens are so much like the description and illustrations of *M. weismanni* that I am certain that they are the same species. I have used Sars' material in the re-description of the species.

Ishikawa (1896) described *M. weismanni* as follows (fig. 23) (Ishikawa, 1896: pp. 1–4):

The length of a middle sized specimen from the tip of the head to the root of the caudal setae measures about 1 mm.

The head rounded and rather short, with a very slight protuberance between the basis of the antenna nearly similar to the European *M. paradoxa* Weismann.

The upper lip shows no structure peculiar to the species. Like the head of *M. rectirostris*. A slight but distinct

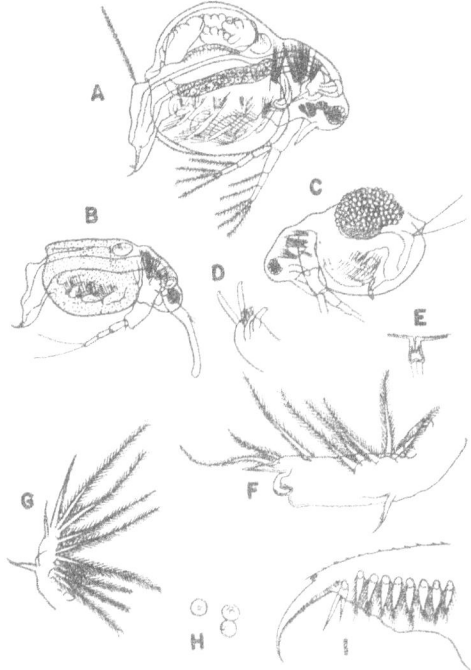

FIG. 23. Ishikawa's illustrations of *Moina weismanni* (after Ishikawa, 1896; illustrated by W. Vars). *A.* Parthenogenetic female. *B.* Male. *C.* Sexual female. *D.* Tip of male antennule. *E.* Posterior tip of shell as seen from above. *F.* First leg of male. *G.* First leg of female. *H.* Spermcells. *I.* Postabdomen of female.

indentation is seen above the eye in front of the coecal processes, but the dorsal border of the head beyond the indentation runs in even curve till to the trunk. This cephalic indentation is shallower in young individuals than in older ones.

The thorax shows no points of difference from other species of the genus. The females with embryoes in the breeding chamber showing the well developed "Nährboden" as usual.

On the dorsal median side of the abdomen there is a well marked process—the "Verschlussfalte"—just as in *Daphnia*, but the fold is not limited only in the median line of the abdomen, but extends on both sides of the trunk in shape of a low side till to the base of the 2nd pair of legs just as Gruber and Weismann first observed both in *M. rectirostris* and *M. paradoxa*. Here also the edge of the shell which came just behind the fold, is greatly thickened, and the breeding chamber is shut up in similar way as in the two European species.

The caudal setae is also long and feathered, the transverse rows of minute setae in the region between the caudal setae and the anus is also found in this species as in the other.

The post-abdomen is provided with eight feathered spines of nearly equal sizes, and with a forked unfeathered one in front just at the base of the apical claw. This bifid spine is much larger than the feathered ones as in *M. rectirostris* (Gruber and Weismann).

The convex side of the apical claw is quite even, except at its base, where a group of sharp spines is to be seen. Its external side is, however, provided with a row of very fine ciliation as in *M. paradoxa*, but the cilia of the basal third of the claw is [*sic*] nearly twice as large as those placed distally, thus approaching more to those of *M. rectirostris*. These basal setae, however, are much more numerous and very much smaller than in that species.

The first pair of antenua [*sic*] is provided with eight olfactory setae, and with a sensory hair at about the median point of its front side. Judging from the figures given by Gruber and Weismann this hair appears to be much longer in our species than in either of the two European species above mentioned, being about half as long as the antenua itself. The entire surface is covered with the number of transverse rows of minute setae.

The second pair of antenua is of usual shape. Its basal joint is provided as usual with transverse rows of setae which are stouter than those of the first antenua. Three sensory setae, two on the outer side of basal joint and the one between the two branches of the antenua as well as a small chitinous prickle close to the latter is also present.

The shape of the upper lip as well as its thick hairs no its lower hinder surface mentioned by Gruber and Weismann is also plainly to be seen. The general shape of the five pairs of legs is in essential point also similar to those of the two other species above mentioned, so that we will only give the description of the first pairs of legs by which our present species can also be easily distinguished from its European allies. In its general shape this leg approaches more to that of *M. rectirostris* than to *M. paradoxa*. The shaft and the four joints can here also be easily distinguished as in the other. The same can be said of the number and the proportional length of the feathered setae, the only difference being the unequal sizes of the two feathered setae of the shaft and the setae of the third and the fourth joints which are not similarly constructed as those of *M. rectirostris*, but more like those of *paradoxa*. In this latter respect the present species stands again midway between these two species.

The entire surface of the shell is punctuated with stellate figures as in many other species.

Just as in *rectirostris* the front and the greater part of the lower margin of the shell is provided with scanty but strong setae. From about the middle portion backwards these setae are replaced by minute sharp teeth which are arranged one after the other in groups of 3–12. There is also a deep indentation at the median posterior edge of the shell, with a pair of strongly bent teeth at its entrance, but differs from those of *M. rectirostris* in the shape and the size of the teeth as well as by the possession of very fine serration on both sides of the indentation.

The ephippium contains only a single egg of oval shape; the longer diameter of which corresponding with that of the animal, the egg lying in either the right or the left side of the animal so long as the ephippium is attached to the body of the mother.

The entire surface of the ephippium is marked with polygonal areas each of which raised up in the shape of a high knob-like protuberance.

The parthenogenetic eggs are as usual spherical and very small, having only 0.1 mm. in diameter. The very scanty deutoplasm is generally of a clear watery colour.

The gamogenetic egg is much larger, of an oval shape, with its long diameter 0.2–0.25 mm. and its short diameter 0.125–0.165 mm. the colour of the fine granular yolk is of a beautiful chalk-white.

The Male

The male is as usual smaller than the female, being only 0.6–0.7 mm. in its longer diameter, but the general shape of the body does not differ much from the other sex, excepting only by having the lower side of the head more straight, and the second pair of antenua comparatively a little longer.

The principal points of difference between the sexes lie as usual in the first antenua and the first pair of legs.

The shape of the first pair of antenua is again like that of *rectirostris*, but the knee like bend on its anterior side occurs more proximally than in that species the number of the olfactory setae at the tip of the antenua appears to be eight or nine, and the hooks only four. These latter are not of equal sizes, two of which being nearly twice as long as the other two.

The general shape of the first pair of legs is as in the other sex, more like to that of *M. rectirostris* than to *M. paradoxa*. Unlike to both these species, there are, however, two setae of unequal size on the shaft of the leg just as in the other sex. The first, and second joint is also similarly constructed as in the female, but the shorter setae of the third joint is changed into a short knob-like process and abutting against the immovable claw, forms a beautiful clasper. The immovable claw formed by the ventral of the three setae of the fourth joint is, however, not curved as in the two European species, but is straight and spine like, so that it does not function in catching the female.

The protoplasm of the small spherical sperm-cells (0.012 mm. in diameter) does not entirely fill up the body of the cell, but is distributed radially around the excentrically placed nucleus.

Female

The head is broadly rounded and very similar to the head of *Moina affinis*. There is a well-developed supraocular depression above the eye. The eye is of a moderate size and is located near the front margin of the head. The antennules are short and are spaced widely apart on the ventral margin of the head. In some instances the antennules may be spread laterally. There is a row of long hairs on the posterior margin of the antennules.

The second antennae are of usual form but are not so pubescent as typical of the genus. The exopod ramus has a vertical row of teeth along its inner margin that are arranged in groups and extend the full length of each segment.

The dorsal surface of the head and shell has a few long hairs as does also the ventral and anterior margins of the shell. They are not as dense as on male specimens of this species.

The ventral shell rim has about seventeen long setae, followed behind by groups of much shorter setae (fig. 25, *B*). There is a pair of elbow-shaped hooks at the dorsal end of the posterior shell margin.

There is a horseshoe-shaped fold lining the posterior part of the body that serves to close off the brood pouch in order to retain the embryos in the pouch. This fold is not as well developed as in *Moina reticulata* or *Moinodaphnia*.

The postabdomen is ornamented behind with the usual pattern of short setae but there are no long hairs on the dorsal margin. There are seven to nine lateral feathered teeth plus a long bident tooth with two long arms. The proximal arm of this bident tooth is much longer than normally found in *Moina*. The claw has a "Basaldorn" of five to seven teeth.

The dorsal base of the claw has a pecten of about twenty thin teeth, and the remaining part of the claw is covered with small setae.

The ephippium has only one sexual egg and is completely covered with raised knobs much as in *affinis*. The ephippium is expanded laterally.

The females are usually less than 1 mm. long.

Male

The head of the male of *weismanni* has a distinct supraocular depression above the eye (fig. 25, *A*). The antennules originate at the tip of the head and just below the eye. These are bent at a point about one-fourth the distance from the head, and the bend is much more distinct than found in *affinis*. The two sensory setae are both located near this bend—one is on the medial margin while the second originates on the lateral margin. The latter seta is longer than the first. The tip of the antennule has four long recurved hooks.

The second antennae are like those of the female.

The head lacks hairs, and no hairs can be seen on the dorsal margin of the shell; but there is a dense cover of hairs on the antero-ventral area of the shell.

There are about nineteen long thin setae on the ventral shell rim; these setae are longer than normally found in the genus. Behind these setae there are several groups of shorter setae that increase in size posteriorly in each group.

The first leg of the male has a weakly developed hook on the third segment. The base of this hook is quite broad, but the remaining part tapers to a point. Of the three setae on the terminal segment, two are feathered while the third forms a thin but short hook. The seta normally arising opposite the hook is greatly reduced in size.

The testes lie along side of the abdomen and extend onto the postabdomen. The precise position of the genital opening could not be seen on the postabdomen, but the vas deferens could be seen on the ventral side of the postabdomen. The opening is probably in a similar position as in *affinis*. The spermatozoa are small spherical cells.

The claw has only a faint pecten of thin hairs. There are six lateral feathered teeth and a long bident tooth that has both arms well developed.

Type material of *Moina brevicornis* is stored in the Zoologisk Museum, Oslo, Norway. Unfortunately, there is no type material of *Moina weismanni* available.

DIFFERENTIAL DIAGNOSIS

Moina weismanni has many affinities with *M. affinis* and *flexuosa*. The female should easily be distinguished from all other species, except *micrura*, by its small size and by the very short antennules. The presence of hairs on part of the head and shell, the pattern of teeth on the second antennae, and the bident tooth with both arms quite long can also be used to characterize the species. The antennules of the female help to distinguish it from *micrura*. The males of *micrura* and *weismanni* differ considerably because males of *weismanni* have the antennules bent near the head, have hairs on the shell, and have a small hook on the first leg. Males of *Moina micrura* have the antennules bent near the middle, completely lack hairs on the shell, and have a well-developed hook on the first leg.

Moina weismanni differs from both *affinis* and *flexuosa* since the setae along the posterior shell margin are grouped while the latter two species have short ungrouped setae. *Moina weismanni* and *Moina affinis* have hairs on both the head and shell; whereas *flexuosa* females lack these hairs altogether.

The antennule of the male *weismanni* has a distinct bend at a point about one-fifth to one-fourth the distance from the head—*affinis* males have the antennules only slightly bent at the base.

DISTRIBUTION

This species has been reported from many parts of the Far East, particularly Japan and China (Ishikawa, 1896; Sars, 1903*a*; fig. 20). Brehm (1951) has identified it from Cambodia and in a later paper (Brehm, 1953*a*) described a form from Mondvi, India (see his *Moina* vom Mondvi fleuve—his *M.* cf. *weismanni* appears instead to be *micrura*) that also appears to be *weismanni*.

Moina weismanni is commonly associated with rice fields and irrigation ditches. This species seems to thrive well in these habitats and one must conclude that it is well adapted for such an environment.

Specimens Examined

China
  Area around Pucheng (Sars' material; no date or precise locality given—determined as *Moina brevicornis*).

MOINA FLEXUOSA SARS, 1897

Moina flexuosa Sars, 1897: pp. 18–23, pl. 3, figs. 1–7.

DIAGNOSIS

One of the small species of *Moina* that never measures over .9 mm. long. The parthenogenetic females have a rather triangular-shaped head with a slight supraocular depression above the eye. The antennules are large and with many rings of setae.

The ventral shell rim is characteristically rounded in front and straight along its posterior half. The front carries about eighteen long setae. Behind these there is a row of equal-sized short setae.

The postabdomen has six to seven lateral feathered teeth and a long bident tooth with both arms long. The dorsal margin of the postabdomen has many long hairs. The claw lacks a pecten.

The ephippium has one egg and is ornamented with knoblike protuberances.

The male antennules are bent near the head and have four terminal hooks. The shell of the male is densely covered with hairs near the ventral and anterior shell rims. The first leg of the male is greatly reduced in size.

DESCRIPTION

Unfortunately, I have seen no specimens of *M. flexuosa* other than Sars' type material, and these were

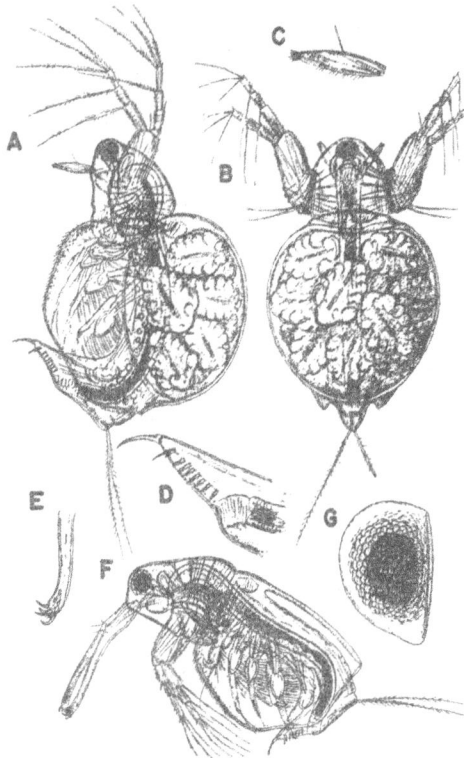

FIG. 24. Sars' original illustrations of *Moina flexuosa* (from Sars, 1897). *A*. Parthenogenetic female. *B*. Dorsal view of same. *C*. Female antennule. *D*. Female postabdomen. *E*. Distal half of male antennule. *F*. Male. *G*. Ephippium.

not well preserved. For this reason, it is very difficult to know how much variability is found within the species. Furthermore, it is necessary to rely heavily on the figures (fig. 24) and description given by Sars which, although they do not always give the desired detail, cannot be questioned for accuracy and beauty.

Sars (1897) described *M. flexuosa* as follows:

The average length of full adult, ovigerous specimens is from 0.7 mm. to 0.8 mm. The largest specimen found in my aquaria measured 0.9 mm., and it is therefore most probable, that this form never exceeds a length of 1 mm. It is accordingly by far the smallest of the 4 as yet known Australian species.

The general form of the body [fig. 24, *A*, *B*] agrees on the whole with that of the other species of the genus, being rather short and stout, with the head and carapace very sharply marked off from each other.

The shell or carapace, which is defined dorsally from the head by a deep depression, has the dorsal part often enormously expanded, in order to make room for the numerous developing young ones. When these are ready to escape from the matrix, that part is sometimes found to project as a large, almost globular pouch sharply defined from the true valvular part, which in all specimens preserves its form quite unaltered. At the junction between the two, the carapace projects posteriorly to a short obtuse prominence, immediately below which there is a slight notch in the hind edges. The ventral edges of the valves, which in all other known species appear almost straight and horizontal, are in the present form rather irregularly flexuous, projecting in the middle to an obtuse prominence, behind which the edges appear slightly concave, before joining the posterior margin. In front of the median prominence the edges are clothed with delicate bristles, and just within the margin a cellular stratum occurs similar to that found in other species. I have, however, failed to detect on the valves any trace of the usual irregular striation.

The head is comparatively small and, seen laterally [fig. 24, *A*], of a somewhat triangular outline. It is remarkably erect, not, as in the other known species, procumbent, and has the frontal part rather prominent and narrowly rounded at the tip, with only a very slight indication of a sinus above. The ventral margin of the head is nearly straight and horizontal, being continuous with the labrum; the dorsal margin appears slightly vaulted. Seen from above [fig. 24, *B*], the head appears rather broad, subpentagonal in form, with the greatest width about equalling the height, and the front obtusely rounded. The fornix is well defined, though not very prominent, and occurs just above the base of the oars.

The eye, occurring just within the frontal part, is not particularly large, exhibiting, however, the usual structure. As in the other species, no trace of an ocellus is to be detected.

The antennulae [fig. 24, *C*] equal about half the length of the head. They are, as in the other species, freely mobile and of a subfusiform shape, with the posterior edge finely ciliated. The sensory bristle of the anterior edge occurs about in the middle. The apical olfactory papillae are very small.

The antennae or oars [fig. 24, *A*, *B*] are powerfully developed, and agree in their structure exactly with those in the other species of the genus. At their base exteriorly 2 remarkable large, juxtaposed setae occur pointing straight outwards.

The tail, which, as usual, does not admit of being wholly withdrawn between the valves, but is constantly seen projecting behind them, exhibits the structure characteristic of the genus. At about the middle, the posterior, or dorsal edge forms a conspicuous bulging, and at this place the anal orifice occurs. The outer part of the tail, beyond the anal orifice [fig. 24, *D*], is conically tapered, and carries on each side a series of about 8 denticles, the outermost of which is placed at some distance from the others and terminates in 2 unequal points. The remaining 7 denticles are very delicate, somewhat flattened, and finely ciliated on both edges. The caudal claws are of moderate length and perfectly smooth, without a trace of secondary denticles. The caudal setae somewhat exceed the tail in length, and are distinctly biarticulate and densely ciliated.

The ephippium, seen laterally [fig. 24, *G*], is of an oval triangular form, being broadly rounded in front, and conically produced behind. It is very coarsely sculptured in the centre with raised knob-like prominences, and always contains but a single egg-ampulla placed longitudinally.

The adult male [fig. 24, *F*] is scarcely more than half as large as the female, and on the whole resemble the males of the 3 other Australian species, though it may be at once distinguished from them by the peculiar bend of the ventral edges of the valves, which exactly agrees with that found in the female. The head is comparatively much larger than in the female, and appears obtusely truncated in front, forming between the bases of the antennules a well-marked obtuse prominence.

The antennulae, as in the males of other species, are very powerfully developed, fully equalling half the length of the body. Their basal part is somewhat thickened, and contains a strong muscular band passing through it diagonally. From the place where this muscle terminates, springs the sensory bristle of the anterior edge, and immediately inside it, a small hook-shaped denticle is seen projecting inwards. The outer part of the antennula is rather narrow and more or less curved, so as to meet, with its tip, the corresponding antennula on the other side, when bent in. Each antennula is armed on the obtusely rounded end with 4 strongly curved hooks, between which the usual fascicle of olfactory papillae may be discerned [fig. 24, *E*].

The structure of the 1st pair of legs and that of the testes seems to agree with that found in *M. australiensis* and *tenuicornis*.

In both sexes the body is highly pellucid and almost colourless. Only in large female specimens, a more or less distinctly rosy tinge may sometimes be observed. The egg contained in the ephippium is of a brick-red colour. (Sars, 1897: pp. 19–22.)

Female

The overall body form and the shape of the ventral shell rim are best characterized by Sars' illustrations which are reproduced here (fig. 24). The flexuose shell rim is readily apparent on Sars' type specimens; however, it is not so noticeable that it completely defines the species. There are other characters that should also be used for this purpose.

The antennules are large and robust, much larger than on any other small species of *Moina* and are as long as the basipod segment of the second antenna, (fig. 25, *C*). The antennules have several rings of short setae. There is also a vertical row of long hairs on the lateral margin of the antennule.

The second antenna is of the usual form and has the vertical row of teeth on the exopod. These teeth extend the full length of each segment, much as in *affinis*. This would help distinguish the species from *M. micrura* for in that species these teeth extend only over the proximal one-half or two-thirds of each segment.

The shell does not seem to be reticulated. The ventral shell rim has about eighteen long setae on the front rounded half, and a row of short, equally sized setae on the straight posterior half of the shell. There is a pair of rounded shell hooks at the dorsal end of the posterior shell margin.

The first leg has the usual setation pattern as in most other species of *Moina*.

The postabdomen is of a typical form but has several rows of long hairs near the dorsal margin (fig. 25, *D*). In addition, there are rows of short setae that are interconnected. The distal conical portion of the postabdomen has one long bident tooth that has both arms rather long although the proximal arm is slightly shorter than the distal arm. There are six or seven lateral feathered teeth. The claw lacks a distinct pecten but is armed with a row of short setae.

The ephippium of the sexual female is expanded laterally and is ornamented with many knoblike protuberances similar to *M. australiensis* and *wierzejskii*.

Male

The male is much smaller than the female. The eye is rather large and nearly fills the tip of the head. The antennules are bent very near the head and have two sensory setae that originate near the knee. According to Sars, one seta is hooklike though this may be an artifact of preservation because these setae are normally not bent. The terminal end of the antennule has four recurved hooks.

The shell of the male is similar to that of the female although not expanded dorsally. According to Sars, the shell has the flexuose ventral margin, but I was unable to see this clearly on the specimens that I examined. However, these specimens were badly distorted because of poor preservation. There were hairs present on the anterior part of the shell as in *M. weismanni*. These are not seen on the head and are completely absent on female specimens.

The hook on the first leg is reduced in size and is pointed at the end. The third segment of the leg is long but has few hairs or setae on the medial margin. The terminal segment has three setae. One is curved and bare, resembling a hook. The other two are feathered. Sars illustrates what appears to be an exopod with a long seta on the male first leg (see fig. 24, *F*). This was not present on the males that I examined.

The structure of the first leg does not agree with that of *australiensis* and *tenuicornis* as stated by Sars (p. 22) since it lacks the exopod and the hook is not so well developed as in either of the other two species.

Rather, the first leg of the male is quite similar to that of males of *affinis*, *weismanni*, and *wierzejskii*.

The testes are similar to those of *australiensis* and *tenuicornis* and presumably the species has spherical spermatozoa. I was unable to see these features on the specimens that I examined. Owing to the close similarity of *flexuosa* in other features to *affinis* and *weismanni*, one would also expect *flexuosa* to have a similar testis structure and spermatozoa which would be in accordance with Sars' statement.

The type material of *M. flexuosa* is in the Zoologisk Museum, Oslo, Norway.

DIFFERENTIAL DIAGNOSIS

Moina flexuosa may be distinguished from most other species of *Moina* by its small size. The only species that might be confused with it are *affinis*, *weismanni*, and perhaps *micrura* and *minuta*. *Moina minuta* has a unique first leg that lacks the anterior seta on the penultimate segment. *Moina micrura* lacks the many rings of setae on the antennule but has groups of short setae along the posterior rim of the shell instead of the short ungrouped setae of *flexuosa*.

Moina flexuosa belongs to the same species group with *affinis* and *weismanni* having many features in common with these two species. However, the parthenogenetic females lack hairs on the head and shells, have only eighteen to twenty ventral shell setae, and have the flexuose ventral shell rim. The antennules are very long—as long as the basipod of the second antennae. The antennules of both *affinis* and *weismanni* are short. The claw lacks a pecten whereas *affinis* has a very distinct pecten. *Moina weismanni* has the setae on the posterior shell margin in groups and of unequal size.

The males of *flexuosa* may easily be distinguished from those of *affinis* because the bend in the antennules is not so near the base. Furthermore, the claws of the postabdomen lack a pecten. On the other hand, the males of *weismanni* are very similar to those of *flexuosa*. Both have hairs near the antero-ventral margins of the shell; both have similar antennules and first trunk limbs. Perhaps the flexuous shell margin of *flexuousa* or the slight pecten on claws of *weismanni* males, as well as the grouped setae on the posterior margin of the shell in the latter will serve to differentiate the males of these two species. However, it is apparent that these two species are very much alike which suggest a close phylogenetic relationship.

DISTRIBUTION

Moina flexuosa has only been reported from Western Australia (fig. 20) by Sars (1897) who cultured specimens hatched from resting eggs contained in dried material (mud) collected "from places where it

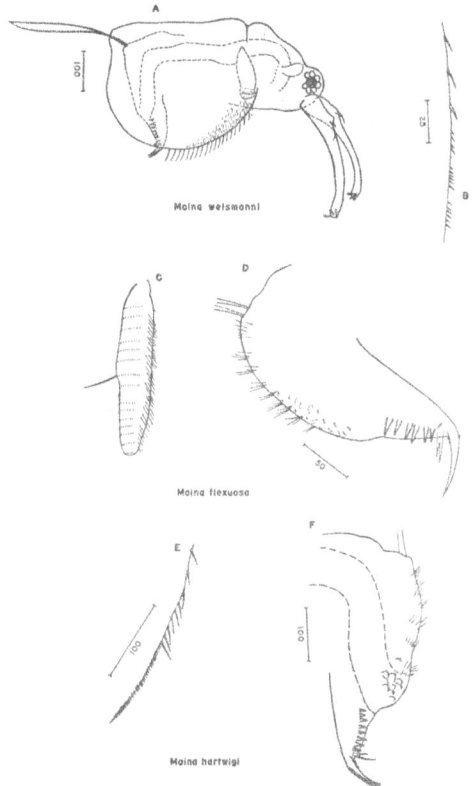

FIG. 25. *Moina weismanni* Ishikawa; *Moina flexuosa* Sars; and *Moina hartwigi* Weltner. *A.* Male of *Moina weismanni* from Pucheng, China (Sars Collection type material of *Moina brevicornis*). *B.* Ventral shell rim of female with long ventral setae and short, grouped setae of posterior rim. Same collection as *A.* *C.* Female antennule of *Moina flexuosa* from type material in Sars Collection. *D.* Female postabdomen of *C.* *E.* Shell rim of *Moina hartwigi* female from type material collected by Stuhlmann. *F.* Female postabdomen of *D.*

was supposed that, at an earlier time of the season, rain-water accumulated" (Sars, 1897: p. 3). Sars did not state the precise locality other than "shallow depressions in the sandy desert, at a distance of about 40 miles east of Roebuck Bay" which is in the northern district of Western Australia. It is probable, however, that this species will be found throughout the large desert areas of Central and Western Australia.

Specimens Examined

Australia:

Western Australia: sand dune area forty miles east of Roebuck Bay (Sars Collection No. 53.2/11).

MOINA HARTWIGI WELTNER, 1898

Moina n. sp. Stuhlmann, 1889 : p. 649.
Moina hartwigi Weltner, 1898 : pp. 135–140 ; 2 figs.
Moina dubia pectinata (partim) Gauthier, 1954 : pp.
   38–41 ; pl. 19, figs. *A–C* ; pl. 20, figs. *A–C* ; pl. 21,
   fig. *G* ; pl. 22, figs. *A*, *B*(?), *D* ; pl. 23, figs. *C*, *D*(?) ;
   pl. 24, figs. *A*, *E*, *I*, *J* ; pl. 25, figs. *A*, *C* ; pl. 26, fig. *A*.

TAXONOMIC NOTES

This species was first collected from Quilimane,
Mozambique, in Eastern Africa by Stuhlmann (1889).
Stuhlmann did not name the species in his published
description but instead gave specimens to Weltner
who, in 1898, published a description and named the
species *hartwigi*. The species has not subsequently
been reliably reported (Verestschagin, 1914, actually
had *M. micrura*) and several investigators have suggested that *hartwigi* is identical to Sars' species *M.
propinqua* (=*micrura*). This, however, is not the
case ; I have examined paratypes of *hartwigi* and have
found it to be a distinct species.

In 1954 Gauthier described a new subspecies of
M. dubia that he called *pectinata* because of the large
claw pecten. This new subspecies was collected from
Madagascar (Malagasy), across the Mozambique
Channel from Quilimane. Gauthier's subspecies
possessed ungrouped setae on the posterior shell rim,
a characteristic of *hartwigi* (see Gauthier, 1954 : pl. 21,
fig. *G*) but not of *M. micrura*. On the basis of this
character as well as others, it seems probable that a
large part of Gauthier's material from Madagascar
was actually *M. hartwigi* (including specimens from
Ikonka, Ambovambe, and perhaps Antanimora).

I have seen only paratype material of *M. hartwigi*
and can, therefore, have no impression of the morphological variability within the species. For this reason,
it is difficult to know precisely how much of the
material of Gauthier's *M. dubia pectinata* should be
ascribed to *M. hartwigi*.

No ephippial females or mature males (one immature male was found) were present in the paratype
material that I examined. The descriptions of the
ephippium and the males are therefore largely based
on comments by Stuhlmann (1889) and Weltner
(1898) or on the description and illustrations given by
Gauthier (1954) which may only in part refer to
hartwigi.

DIAGNOSIS

The females of *Moina hartwigi* are about 1.1 mm.
long and have a well-developed supraocular depression
on the head. There are no hairs on the shell or dorsal
margin of the head, but there is a patch of hairs on the
ventral head margin behind the antennules. The
posterior margin of the shell has a row of ungrouped
setae. The postabdomen has long hairs on the dorsal

margin, and the claw has a pecten of about fifteen thin
teeth. Only one sexual egg is produced.

The males are known only from a brief description
by Stuhlmann (1889). They are supposedly very
much like the males of *micrura* and have the same type
of spermatozoa, i.e., small round cells with many
radiating axons.

DESCRIPTION

I have included the descriptions of both Stuhlmann
(1889) and Weltner (1898) because the former mentioned characters not referred to by Weltner. Stuhlmann's remarks also included a brief description of the
collecting locality which was described as a fish tank
from which he had collected *Protopterus*.

The following is Stuhlmann's description :

> Dieselbe unterscheidet sich von der nahe verwandten
> *M. micrura* durch den Besitz von 8–9 bewimperten,
> kegelförmigen Dornen am Postabdomen ; die Endkralle
> trägt einen Nebenkamm und dorsalwärts etwa 6 Neben-
> dornen. Das Thier ist hellgelbrötlich, besonders in
> Herzgegend und Nährboden, auch manche farbige Fett-
> tropfen tragen zur Färbung bei. Das Ephippium beher-
> bergt ein Ei. Das bedeutend kleinere Männchen zeichnet
> sich durch längere (etwas weniger als halbe Körperlänge)
> Tastantennen aus, die am Ende mit zwei dem Körper
> zugewandten Klauen bewehrt sind. Das erste Beinpaar
> trägt einen mässig grossen Haken. Die Form der Samen-
> körperchen liesse sich am besten mit der von *Actinophrys*
> vergleichen. (Stuhlmann, 1889 : p. 649.)

Weltner's description was much more detailed but
did not necessarily include other important species
characters (fig. 26).

> Die Körperform des mit zahlreichen Embryonen
> erfüllten Weibchens zeigt grosse Ähnlichkeit mit der
> australischen M. propinqua Sars, nur ist der Kopf nicht
> wie ihr niedergedrückt sondern hoch, und die hintere
> Kopfkante ist bei hartwigi in der Regel nicht in ihrer
> ganzen Länge convex, sondern verläuft im mittleren Teile
> mehr gerade. Bei einzelnen Exemplaren war jedoch der
> hintere Kopfrand ganz convex und auch Stuhlmann hat
> dies beobachtet, wie eine seiner Zeichnungen beweist.
> Unterhalb des Auges ist der Kopf eingebuchtet wie bei
> propinqua und ist wie bei dieser vom Rumpfe durch einen
> tiefen Einschnitt abgesetzt. Die Stirn ist gerundet und

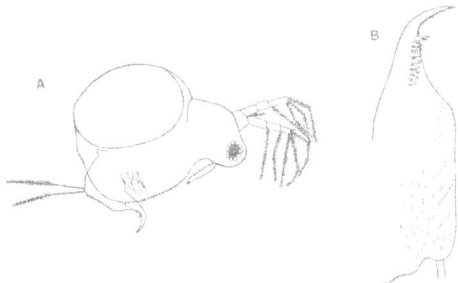

FIG. 26.   Weltner's illustrations of *Moina hartwigi* (from Weltner,
1898).   *A*. Parthenogenetic female.   *B*. Postabdomen.

bildet den vordersten Teil des Kopfes. Der Unterrand desselben ist unterhalb der ersten Antenne eingebogen und zieht dann nach vorne zum unteren Schalenrande hin. Das grosse Auge liegt im vorderen Teile des Kopfes, im optischen Durchschnitt zähle ich 7–9 stark hervortretende Linsen. Ein Nebenauge fehlt wie bei den anderen Arten der Gattung.

Der Rumpf ist bei den trächtigen Weibchen sehr viel breiter als der Kopf und gewinnt vom Rüchen oder Bauch gesehen das Ansehen einer Kugel (wie propinqua). Der vordere Rand der Schale ist wie bei dieser etwas ausgebuchtet und lässt heir einen Teil der Mandibeln frei. Der ventrale Rand ist fast gerade oder nur sehr wenig convex und vorne mit Borsten, im hinteren Teile mit Zähnchen besetzt in der Weise, wie das bei vielen Arten dieses Genus vorkommt (aber bei propinqua nicht der Fall zu sein scheint). Der hintere Schalenrand lässt einen stumpfen, breiten Lappen ganz ähnlich dem von propinqua erkennen. Die Schalenskulptur besteht aus quer über die Schalen hinziehenden Linien, die hier und da, besonders am ventralen Rande anastomosieren, so dass an solchen Stellen die Schale unregelmässig gefeldert erscheint. Die Entfernung der Querlinien von einander beträgt 0,008– 0,012 mm. Eine solche Skulptur wird bei den Arten der Gattung Moina nur von lilljeborgi Schödl., propinqua Sars und affinis Birge erwähnt. Auf der nebenstehenden Figur habe ich ein Stück dieser Skulptur von M. hartwigi wiedergegeben.

Die ersten Antennen sind gerade und entweder in der Mitte etwas verdünnt oder es ist nur der äussere Rand in der Mitte eingebogen, vor dieser Einsenkung steht die Sinnesborste. Der innere Rand ist lang behaart. Die Sinnesborsten sind kurz, ihre Anzahl beträgt 6–8. Jeder Stamm der zweiten Antennen trägt an seiner Basis auf der Ventralseite zwei Borsten, deren jede zweigliedrig ist. Das zweite Glied dieser Borsten ist 3 bis 4 mal so lang als das erste und allseitig lang behaart. Die Borste, welche zwischen den beiden Ästen jeder Antenne steht, ist ebenfalls zweigliedrig mit längerem allseitig behartem zweitem Gliede. Die Borsten der Ruderantennenäste sind zweigliedrig, das zweite Glied länger als das erste, beide Glieder sind allseitig lang behaart.

Das Abdomen ist bei den meisten Exemplaren wie bei propinqua nicht ganz eingezogen: der Krallenteil meist aus der Schale hervor. Andere Stücke haben ihr Abdomen ganz retrahiert. Dieses trägt dorsal eine kurze konische Falte, mit welcher der Brutraum zum Teil abgeschlossen wird, wie bei propinqua. Die Stelle, an der die Analöffnung liegt, ist vorgezogen. Die dorsale Kante des Abdomens verläuft ziemlich gerade und ist stets mit kurzen Borsten versehen, die zum Teil einzeln, zum Teil in Gruppen zu zwei bis drei stehen. Ein Blick von oben auf das Abdomen lehrt, dass die Borsten nach dem Anus hin in zwei Reihen und nach den beiden setae hin nur in einer Reihe stehen. Ich zähle in seitlicher Lage des Abdomens zehn bis zwölf solcher Borsten resp. Borstengruppen. Eine ähnliche Bewehrung scheint nach der Abbildung zu urteilen, bei M. affinis vorhanden zu sein, mit der aber unsere Art nicht identisch ist. Bei M. hartwigi sind auch die Seitenteile des Abdomens bewehrt und zwar bei den meisten Exemplaren in der von Richard bei M. dubia beschriebenen Weise mit Querreihen kurzer Stacheln. Ausserdem finden sich aber stets noch zerstreut stehende Borsten an den Seitenteilen in grösserer oder geringerer Anzahl. Soviel aus der Litteratur zu ersehen ist, besitazen folgende Arten ähnlich bewehrte Abdomina: M. rectirostris bei Daday, 1888, Taf. 3, Fig. 3, M. salina das Fig. 4, M. dubia Richard 1895 fig. und M wierzejski Richard 1895 fig., jedoch fehlen diesen Arten die an den Seitenteilen stehenden zerstreuten Borsten und die borsten

am dorsalen Rande des Abdomens, weiter entbehren sie die noch zu erwähnenden Zähnchenreihen von der Endklaue. Auch unterscheiden sich die genannten Arten in anderer Beziehung von M. hartwigi. Bei dieser finden sich am distalen Ende des Abdomens 7–9 in der gewöhnlichen Weise bewimperte Zähne und ein zweigespaltener Zahn, dessen hinterer länger als der vordere ist. Auf der ventralen Seite des Abdomens bemerkt man dicht vor der Endkralle 4–9 quer verlaufende feine Zahnreihen, auf der Kralle selbst 3–5 dorsale Zähne und unten einen Nebenkamm, der aus 12–15 Zähnen besteht; der übrige Teil der Klauen ist fein bezahnt. Die Abdominalborsten sind wie gewöhnlich lang, zweigliederig, das erste Glied ist meist kürzer als das zweite behaarte.

Ephippientragende ♀ habe ich in dem Material nicht gefunden; nach Stuhlmann trägt jedes Ephippium nur ein Ei.

Das Männchen gleicht im Habitus ganz dem M. propinqua Sars. Besonders fällt an ihm gegenüber dem Weibchen der hohe Kopf auf. Wie bei propinqua sind die ersten Antennen sehr lang und erreichen die Länge des Kopfes; ihr äusserer Bau weicht nicht von der Schilderung, die Sars von seiner propinqua giebt, ab. Nach Stuhlmann l. c. besitzen diese Antennen zwei Endklauen, dagegen fand ich an dem einzigen von mir beobachteten ♂ deren drei; darnach ist die Zahl der Endhaken auf 2 bis 3 anzugeben. Den Bau des ersten Beinpaares habe ich nicht untersucht, weil Stuhlmanns Zeichnung erkennen lässt, dass auch hier 3 Borsten und ein Haken vorhanden sind, letzterer ist kleiner als der bei propinqua. Das männliche Abdomen ist wie beim ♀ beschaffen und ebenfalls am dorsalen Rande und an den Seiten mit den Borsten bewehrt.

Länge des ♀ : 0.98 bis 1,12, Länge des ♂ : 0,77 mm. (Weltner, 1898: pp. 137–139.)

Female

The head is rounded in front and there is a distinct supraocular depression (fig. 26, A). The eye is large and lies near the anterior margin. There is a small patch of hairs behind the antennules on the ventral side of the head. This is similar to the pattern of hairs on the head of M. brachiata. The antennules are long and thin. There is a vertical row of hairs along the lateral margin, but there do not appear to be rings of short setae as found in M. brachiata. The labrum is of the usual form and may have a few hairs on its ventral margin.

The second antennae are very pubescent. The exopod ramus carries an inner row of teeth that extend over the proximal part of each segment but do not extend to the distal end.

The shell lacks surface hairs and has a distinct pattern of reticulations. The ventral shell rim has from twenty-three to twenty-seven long setae while the posterior rim has only short setae that are of equal size and are not arranged in groups (fig. 25, E). The dorsal end of the posterior rim has a pair of curved shell hooks.

The first trunk limb has both anterior setae that are armed with fine hairs. The limb is similar to the typical moinid limb.

The fifth limb is almost identical to the limb of brachiata.

There is an abdominal fold that defines the posterior edge of the brood pouch. The postabdomen has long hairs on the dorsal rim in addition to the short setae usually found on the postabdomen of *Moina*. The distal conical part of the postabdomen has seven to eight lateral feathered teeth and a long bident tooth that has the proximal arm about one-half the length of the distal arm (fig. 25, *F*). There is a "Basaldorn" on the ventral side of the claw of about six teeth. The dorsal base of the claw has a pecten of about fifteen thin, but rather long, teeth.

The ephippium of the sexual female contains only one egg. According to Gauthier's (1954: pl. 11, figs. *E* and *G*) illustration of the ephippium of *M. dubia pectinata*, it is almost completely reticulated except for the posterior part that is only indistinctly reticulated. There does not appear to be an embossed sphere as in *brachiata*.

The females are .77 to 1.24 mm. long.

Males

The males are not well known. If Gauthier's (1954) Madagascar specimens are truly *hartwigi* then the males are very similar to those of *Moina micrura* but are slightly larger (compare Gauthier's illustrations on plates 8 and 25). The head is rounded in front and has a distinct supraocular depression. The eye is rather large and fills the anterior end of the head. The antennules are bent at a point about one-third the distance from the proximal end and have two sensory setae at this bend. The distal end has three hooks. This description agrees with that given by Stuhlmann (1889) although he found only two hooks on the antennules. Weltner, however, found three. It is probable that Stuhlmann had examined an immature specimen.

The first leg of the male has a large hook and a rather small distal segment. There is no exopod segment. The first leg as illustrated by Gauthier (1954: pl. 26, figs. *A*, *B*, *C*) is rather similar to the first leg of *micrura*.

The postabdomen of the male is like that of the female although I cannot tell from Gauthier's illustrations whether a pecten is present, nor is it possible to recognize the position of the genital opening.

According to Stuhlmann, the spermatozoa are like "*Actinophrys*" in form. This would mean that the spermatozoa of *hartwigi* are identical to those of *micrura* and *brachiata*.

The males are from .69 to .85 mm. long.

Paratype material of *Moina hartwigi* is preserved in the Zoologisches Museum, Humboldt University, Berlin, East Zone; and the Zoologisches Museum, Hamburg University, Hamburg, Germany.

DIFFERENTIAL DIAGNOSIS

The females of *Moina hartwigi* might be confused with *affinis*, *weismanni*, *brachiata*, or *micrura*, but more commonly with the latter two because *affinis* and *weismanni* have not been reported from Africa to which *hartwigi* is restricted. The patch of hairs on the ventral surface of the head, the ungrouped setae on the posterior shell rim, the hairs on the dorsal margin of the postabdomen, and the pecten on the claws are all characters shared with *affinis* and to some extent with *weismanni*. In *affinis*, however, the hairs normally cover the body while on *weismanni* the posterior shell rim setae are always grouped. Moreover, the males and ephippial females of *hartwigi* are completely different from *affinis* or *weismanni* males and ephippial females, and clearly associate *hartwigi* with both *brachiata* and *micrura*.

Moina brachiata also has the patch of hairs on the head as found on *hartwigi*, but it lacks long hairs on the postabdomen. The claw pecten of *brachiata* is much larger than the pecten of *hartwigi* claws.

Moina micrura completely lacks hairs on the head or shell. The setae on the posterior shell rim are clearly grouped and the Old World forms of *micrura* lack hairs on the dorsal margin of the postabdomen. *Moina micrura ciliata* from South America does have these hairs. The males of *hartwigi* are distinguished from *brachiata* males in having only three distal hooks on the antennule. In this regard, the males are similar to *micrura*. In fact, there seems to be little besides size to distinguish the males of these two species. However, the males of *hartwigi* are not well known, and it is probable that other distinguishing characters will later be found.

Moina hartwigi is closely related to *micrura* and *brachiata* not only by the above features but also by the spermatozoa that in all three forms have a heliozoan shape. Possibly *hartwigi* is more closely associated with *micrura*, especially with regard to the males; but the larger claw pecten and the patch of hairs on the head are strongly reminiscent of *brachiata*. *Moina hartwigi* clearly appears to be an intermediate form if not an ancestral form to these two species.

DISTRIBUTION

Type locality: A fish tank near Quilimane, Mozambique (Stuhlmann, 1889).

Besides Mozambique, this species may also be found on the island of Madagascar (Malagasy). The species probably occurs in other areas of Eastern Africa (fig. 31).

Specimens Examined

Africa:

Mozambique: Quilimane, collected by Stuhlmann, 1889 (paratype material).

MOINA MINUTA HANSEN, 1899

Moina minuta Hansen, 1899: pp. 6–7, pl. 1, figs. 3–3a.
Moinodaphnia brasiliensis Stingelin, 1904: pp. 580–581, pl. 20, figs. 3–4.
Moina minima Spandl, 1926: p. 103, figs. 4a–d.

TAXONOMIC NOTES

Unfortunately there is no type material for any of the described forms belonging to this species. I do have material of a *Moina* from Central America (Guatemala), from Lago Izabal and from the Rio Dulce, a river connecting Izabal with the El Golfete and the Gulf of Honduras. These specimens share the following features with the above mentioned species: (1) they are all reported from lakes or rivers from the east coast of South and Central America; (2) they are all small forms, .5 to .7 mm. long; (3) they are all characterized by having a short swimming seta on the first endopod segment of the second antenna; and (4) all forms have a rather pronounced pecten on the postabdominal claws. The descriptions of the individual forms differ from the specimens in my possession by the following morphological features: (1) *brasiliensis* is reported to have an ocellus; (2) *minuta* is supposed to have a very small compound eye; (3) *minima* was described as with only two distal swimming hairs on the second antennae rami and with the ventral shell rim lined with fine setae or hairs between each large seta. Only the ocellus of *brasiliensis* can be considered an important character. The features that distinguish *minima* do not fit the Moinidae whatsoever and are probably artifacts of poor preservation or due to bad optics because this form, as illustrated, is obviously a moinid. The small eye of *minuta* and the presence or absence of an ocellus would depend upon the state of preservation of the specimens.

One further point that supports the idea that at least two of these forms belong to the same species is that *Moina minuta* and *Moinodaphnia brasiliensis* were both reported from rivers at the mouth of the Amazon River in Brazil. Although I do not have extensive collections of *Moina* from South America, there is no reason to believe that these three described forms, that are clearly not synonyms of *Moina micrura*, should be distinct from one another. This conclusion is based on the similarity in habitat as well as body morphology. Furthermore, since these same features are shared by the Central American specimens from Lago Izabal and the Rio Dulce, I believe that these specimens are also the same form.

The description below is based upon specimens collected from Lago Izabal, Guatemala.

DIAGNOSIS

A small species that measures from .5 to .7 mm. long. Head with a supraocular depression. The eye is large and lies contiguous to the anterior and ventral margins of the head. An ocellus may be present. The antennules are short and have one very long sensory seta that originates from the anterior margin at the mid-point of the antennule. The sensory papillae of the antennules are very long.

The second antennae are of the usual form, but the two sensory setae located just behind the basipod, and the one seta found between the exo- and endopods are very long. The swimming seta on endopod segment one is short.

The shell is not sculptured. The shell rim has a row of thirteen to sixteen marginal setae followed behind by groups of unequally sized short setae.

The first leg of *minuta* lacks the anterior seta on the third segment, and the anterior seta of the ultimate segment is reduced to a small spine.

The postabdomen has three to six lateral feathered teeth and a long bident tooth. The claw is armed with a pecten of eleven to fourteen sharply pointed teeth.

Males and ephippial females are unknown.

The following is Hansen's (1899) description of *Moina minuta* (fig. 27):

Weibchen. Der Kopf nimmt fast ein Drittel der ganzen Läng des Körpers ein; von der Seite gesehen geht sein Oberrand in einer etwas gebogenen Linie vorwärts und abwärts bis zu dem bogenförmigen Vorderrande; sein Unterrand und Labrum bilden einen stumpfen Winkel miteinander. Das Auge ist klein und sitzt sowohl vom Vorderrand des Kopfes als von dessen Unterrand weit entfernt. Die Antennulen messen wenig über ein Drittel der Länge des Kopfes, sie sind etwas spindelförmig, an der Mitte des Vorderrandes mit einer Borste versehen. Der Schaft der Antennen ist schlank; die Borste an der Spitze des ersten Gliedes des 3 gliedrigen Aeste ist kürzer, oder höchstens von derselben Länge wie die 2 letzen Glieder zusammen, während die 4 andern Borsten an jedem der zwei Aeste besonders lang und länger als der längste der Aeste sind. An der Rückenseite ist der Kopf durch eine tiefe Einsenkung vom Rumpf abgesetzt, und die erste kurze Theil der Rückenseite des Rumpfes ist ziemlich stark gewölbt und zugleich durch eine ansehnliche Einsenkung von der über der Bruthöhle liegenden Abschnitt Abgesetzt. Der Hinterrand der Schale ist sehr lang und stark ausgebogen, sodass die Schale das Abdomen des Thieres bedeckt, von dem fast nur die terminalen Fortsätze frei hervorragen. Der unterrand der Schale ist auf einer ziemlich weiten Strecke nur schwach gebogen. Der untere Theil des Vorderrandes der Schale und die vorderste Hälfte des schwach bogenförmigen Unterrandes sind mit einigen kurzen und ziemlich weit von einander sitzenden Borsten ausgestattet. Das Abdomen hat zwischen dem Anus und den terminalen Fortsätzen an jeder Seite 5 Fortsätze [fig. 27, *B*], von denen die 4 proximalen kegelförmig und behaart sind, während der distale nackt und doppelt, oder mehr als doppelt so lang wie der nächstäusserste ist und in zwei Aesten endigt, von denen der eine kurz, der andere besonders lang ist. Dicht über dem obersten Processus war es möglich auf einem einzelnen Exemplar 3 Borsten zu sehen. Die distale Hälfte der terminalen Fortsätze hat an der Aussenseite dicht bei dem Hinterrande eine Reihe äusserst kleiner, länglicher Zähne. Die Abdominalborsten messen ein wenig über die Hälfte

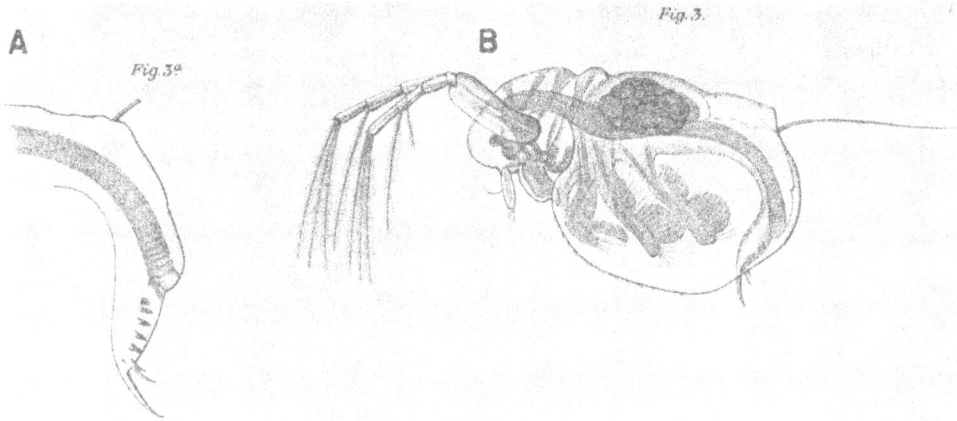

FIG. 27. Hansen's illustrations of *Moina minuta* (from Hansen, 1899). *A*. Parthenogenetic female. *B*. Postabdomen.

der Länge des Körpers. In der Bruthöle 1 Embryo oder 2 Embryonen.—Das grösste Exemplar ist 0,46 mm lang. (Hansen, 1899: pp. 6–7.)

Female

The head is rounded above with a slight protruded area around the eye at the angle of the anterior and ventral margins (fig. 28, *A*). There is a shallow supraocular depression located dorsal to this prominence and just above the plane of the eye. The antennules are set well back along the ventral margin of the head, and each antennule originates from a small knob. The antennules are short and thin with a row of long hairs along the posterior margin. There is a very long sensory seta at the midpoint on the anterior margin of the antennule. The sensory papillae of the antennules are very long in comparison to most other species of *Moina*.

The eye is large and lies contiguous to the anterior and ventral margins of the head. It is composed of a large pigment spot with many small crystalline lenses around it. An ocellus may be present in this species, but I have never seen one.

The second antennae are not well developed; the long swimming setae just reach the posterior margin of the shell. However, the two sensory setae lying at the base of the antennae and the one seta located between the endo- and exopod on the basipod are very long, much longer than in any other species of this genus except for *Moina micrura dubia* (fig. 28, *B*). One of the two basal setae is longer than the basipod segment and first segment of the exopod combined. The seta lying between the exo- and endopod is as long as either ramus.

The swimming seta on the first segment of the endopod is much shorter than the other swimming

setae. The second antenna is not nearly so pubescent as in most other species of *Moina*.

The head is short and comprises less than one-third of the total length of the animal. The shell is very

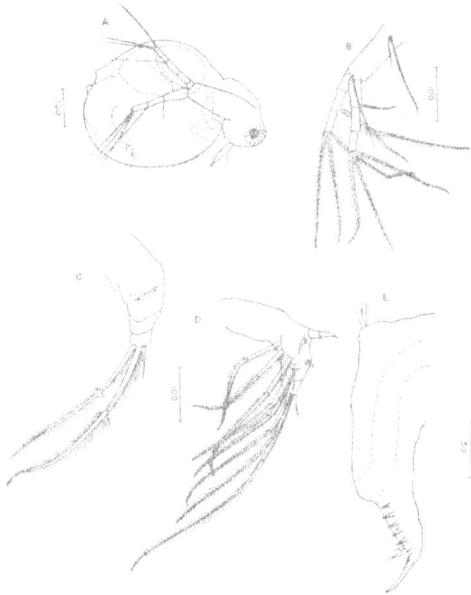

FIG. 28. *Moina minuta* Hansen. *A*. Parthenogenetic female from Lago Izabal, Guatemala. 15-III-05 (Birge Collection). *B*. Second antenna, female from Lago Izabal, Guatemala. 10-VI-64 (collected by M. Tsukada). *C*. Lateral view of first leg of *B*. *D*. Medial view of first leg of *B*. *E*. Postabdomen of *B*.

rotund in outline and lacks surface sculpturing. The ventral shell rim has a row of thirteen to sixteen setae that are followed behind by several groups of shorter setae. Each of these groups consists of eight to ten short setae that increase in size posteriorly. There is a pair of shell hooks, one on each valve, at the point where the shell comes together at the dorsal end of the posterior margin.

The first leg lacks the anterior seta of the penultimate segment (fig. 28, *C* and *D*). The anterior seta of the terminal segment is reduced to a spine and is not feathered. The two other setae of the terminal segment are both feathered.

The postabdomen is long and thin (fig. 28, *E*). The dorsal margin is ornamented with the wavy lines of short setae but there are no long hairs as in *M. micrura ciliata* and *affinis*. The distal conical part of the postabdomen has from three to six lateral feathered teeth that may vary in number on either side of the same postabdomen. The bident tooth is long; the distal arm is about twice as long as the proximal arm. The claw base has a "Basaldorn" of two to three thin teeth and a pecten of twelve to fourteen sharply pointed teeth. The distal part of the claw is armed with a row of fine setae.

No ephippial (sexual) females or males have been found.

The parthenogenetic females that I have measured are from .5 to .7 mm. long.

Specimens of *Moina minuta* from Lago Izabal have been placed in the United States National Museum and the British Museum (Natural History).

DIFFERENTIAL DIAGNOSIS

Unfortunately the ephippial females and males of this species are presently unknown, but there can be no doubt that *M. minuta* is a distinct species owing to the very characteristic first leg and the long sensory setae of the second antennae. These characters serve to differentiate the species from the only two species with which it may be confused in the New World, *Moina micrura* and *affinis*. The form of the head of *minuta* is also quite distinct as is also the proximity of the large eye to the margin. In both *affinis* and *micrura* the eye is set back in the head. Moreover, neither of these species has long sensory setae on the second antennae.

The number of feathered teeth on the postabdomen (three to six) in *minuta* is similar to that found in *micrura* (three to ten), but the pecten on the claw of *minuta* is much larger than that found on all New World forms of *micrura*. However, this pecten is similar to that found on the claws of *affinis*, but at the same time the latter species has six to twelve lateral teeth on the postabdomen and has long hairs on the dorsal margin. *Moina minuta* completely lacks long hairs on the head and shell or on the postabdomen.

Moinodaphnia macleayi has a rather similar body morphology as *minuta*; but in the former species, the postabdomen is of a different form, the claw lacks a pecten, and the bident tooth is much larger. Moreover, *Moinodaphnia macleayi* has a different first leg that has retained the anterior seta on the penultimate segment.

DISTRIBUTION

Type locality: Mouth of the Rio Tocantins, Brazil.

The species has been reported from the Rio Arama, an arm of the Amazon River (Stingelin, 1904), and from Paranagua, Brazil—a large gulf opening to the ocean (Spandl, 1926).

Specimens were found in the Birge Collection labeled as *Moina*, L. Izabal, Guatemala, Mch. (i.e. March) 15, 1905. There was no mention on the labels of who collected this material, and I can find no record that Birge ever described any Cladocera from Central America. Juday's (1916) collections from Guatemala were made in 1910. There were also a few specimens from the Rio Dulce, a river that connects Izabal with the El Golfete and the Gulf of Honduras. These specimens were dated Mch. 19, 1905.

Additional specimens of this species were collected from the open water and littoral areas of Izabal by Matsuo Tsukada on June 10, 1964, and by Allan Covich on July 10, 1965. The species was very common in the littoral collections. I have not found this species in any of the many other collections that I have examined from South or Central America. The species appears to be a limnoplanktonic and littoral form that prefers saline or, at least, oligohaline lakes and rivers. In all instances the species has been found in rivers or lakes opening to the ocean (fig. 31).

Specimens Examined

Central America:
Guatemala:
  Lago de Izabel; collected 15-III-05 (Birge Collection).
  Lago de Izabal; collected 10-VI-64 by M. Tsukada.
  Lago de Izabal; collected 10-VII-65 by A. Covich.
  Rio Dulce; collected 19-III-05 (Birge Collection).

MOINA MONGOLICA DADAY, 1901

Moina mongolica Daday, 1901: pp. 444–445, pl. 21, figs. 11–15, pl. 22, figs. 1–3.
Moina microphthalma Sars, 1903a: pp. 179–180; pl. 7, figs. 6, 6a.
Moina salinarum Gurney, 1909: pp. 289–290; pl. 10, figs. 13–15.
Moina salinarum Gurney, 1911: pp. 28–29; pl. 2, figs. 3–4.
Moina microphthalma Behning, 1941: pp. 165–166; figs. 67a–b.

Moina salinarum Gauthier, 1954: pp. 42–50; pl. 9, figs. *G–H*; pl. 25, figs. *E–F*; pl. 26, figs. *D–E*; pl. 27, figs. *A–E*; pl. 28, figs. *A–D*; pl. 29, figs. *A–F*; pl. 30, figs. *A–J*; pl. 31, figs. *A–E*, and figs. 3–4.
Moina salinarum Löffler, 1961: p. 369; fig. 11.

(?) **Moina salina** Daday, 1888

Daphnia brachiata Stepanov, 1885: p. 30.
Daphnia rectirostris var. *salina* Stepanov, 1886: pp. 197–198.
Moina salina Daday, 1888: p. 112; pl. 3, figs. 5–9.

TAXONOMIC NOTES

The systematic position of this species is not well established, and I have some reservations about the use of the specific name *mongolica*. This form was first properly described by Daday (1901) from Mongolia. I have used Daday's type material in the present description. I have examined a male specimen of the same species identified by Sars as *Moina microphthalma*. The specimen had been collected from Birket Qârûn in Egypt, a locality from which Gurney (1911) reported his previously described species *Moina salinarum* (Gurney 1909). It is apparent from the description of *salinarum* by Gurney (1909, 1911) that it is the same as *Moina mongolica*. Unfortunately, Sars' (1903a) description of *microphthalma* is not complete because it does not include a description of the males or ephippial females. However, I have examined female specimens from the Aral Sea (sent to me by F. D. Mordukhai-Boltovskoi) that are identical to *M. mongolica* and to Sars' description of *microphthalma*.

All of these forms were reported from saline lakes and pools: *mongolica* from Cherman-tzagan-nor (Terhin Tsagen Nûr), Mongolia; *microphthalma* from Lake Tenise near Omsk, U.S.S.R.; and *salinarum* from various lakes in Tunisia, Algeria, and Egypt.

There is an almost continuous zone of saline lakes and pools that extends from North Africa, through the Middle East, South and Central U.S.S.R., and into Mongolia and Manchuria. The same species of *Moina* appears to inhabit these pools.

The first reports of *Moina* from saline pools in Southern Russia were made by Stepanov (1885, 1886) from lakes near Slavyansk in the Ukrainian S.S.R. Stepanov identified this form as *Daphnia brachiata* (1885) but the following year reidentified it as *Daphnia rectirostris* var. *salina*. He did not describe the form on either occasion nor did he give an illustration. Possibly Stepanov did collect the present species since it occurs widely in this region (Behning, 1941), but we have no way of verifying this contention. In any event, Stepanov's comments do not fulfill the requirements for a valid description.

In 1888 Daday described a form from a saline lake in Hungary to which he gave the name "*Moina salina*

Stepanov." Daday included a Latin and a Hungarian description of the males as well as figures of both male and female specimens. His text and figures suggest that this form may be the same as *Moina mongolica* (no type material is available). The figure of the postabdomen is very similar, and the male antennules figured could only be of either *brachiata* or *mongolica*. He illustrated five short terminal hooks on the male antennule (*micrura* has three or four long hooks); the claw lacked a pecten (*brachiata* has a very large claw pecten). The male first leg was very much like that of both *mongolica* and *brachiata*.

However, Daday gave the length of the female as .75 to .90 mm. and the male as .45 to .50 mm. These measurements are much less than those usually recorded for *mongolica* or *salinarum*. Daday (1901) gave the length of females of *mongolica* as .8 to 1.2 mm., but the specimens that I have measured are 1.06 to 1.55 mm. long, and the males measure .91 to 1.00 mm. Gauthier's (1954) females of *salinarum* were 1.09 to 1.80 mm. and males were .83 to .89 mm. long. Behning's (1941) measurements for *M. microphthalma* were 1.0 mm. for the females (probably taken from Sars, 1903) and .8 mm. for the males.

The measurements for the latter three forms are within a plausible range of one another. The measurements for *M. salina* by Daday are not. It is difficult to believe that the males of the same species should vary in size from .4 to 1.0 mm. long. Possibly Daday's measurements for *salina* are wrong (he gives reasonable measurements for his other species in the same paper). In that event it could be assumed that *Moina salina sensu* Daday is the same species as *Moina mongolica*, and that *salina* is therefore the valid name. It may also be assumed that *salina* is an aberrant form or a distinct subspecies.

Without specimens of *Moina salina* or even valid reports of the present species from any part of Europe, including Hungary, I am not convinced that we can make either of these assumptions. I believe it best to use the oldest name known to have been associated with the Old World halophilic species of *Moina*, *Moina mongolica*. If specimens are later found in Hungary, and if female and male specimens are found to measure as little as the lengths given by Daday (1888), then it may be necessary to change the specific name. Until then I believe that the name *Moina salina* should be added to the present synonymy as (?) *salina*.

DIAGNOSIS

The head is rounded and only occasionally has a supraocular depression. The eye is small, the antennules are short and thick. The second antenna lacks the inner row of teeth on the exopod ramus. The shell is rounded, with about twenty-nine long setae on the ventral rim and a row of ungrouped setae on the posterior rim. Shell hooks are absent. The first leg lacks an anterior seta on the third segment. The postabdomen has seven to nine short feathered teeth and a small, thin bident tooth that has both arms equally long. The claw is armed with its pecten of fine hairs. The ephippium has one egg and is reticulated over its entire surface.

The male antennules are long. The short sensory seta originates from the medial margin about one-fourth the distance from the head. The longer seta originates from the anterior margin at a point about one-third the distance from the head. The distal end of antennules has four to five hooks. The first leg lacks an exopod but has a well-developed hook. The terminal segment of the first leg has a small curved seta and two short feathered setae. The spermatozoa are spherical cells.

DESCRIPTION

Daday (1901) described *mongolica* as follows (fig. 29):

♀. Körper gedrungen. Kopf und Rumpf deutlich abgesondert, letzterer höher, doppelt so lang. Der obere Rand des Kopfes stark gewölbt, Stirne deutlich abgerundet, ohne Vertiefung ober dem Auge. Bauchrand des Kopfes in der Mitte stark gebuchtet, hinter der Antennenbasis zugespitzt.

Die 1. Antennen sind ziemlich kurz, spindelförmig, beborstet. Das 2. Antennenpaar ist von jenem der übrigen Arten dieser Gattung nicht verschieden. Das ziemlich kleine Auge liegt am unteren Theil der Stirn; sein Pigment ist kugelförmig.

Schale sehr biegsam, glatt; der Dorsalrand ist bei jüngeren Exemplaren nur sehr wenig, bei älteren, welche ihre Eier bereits ablegten, ziemlich stark gewölbt, bildet aber mit dem Hinterrand stets einen mehr oder minder

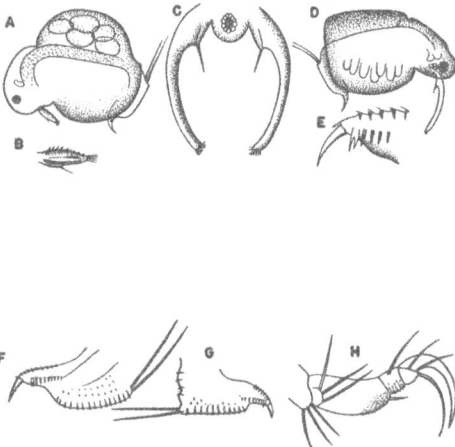

FIG. 29. Daday's illustrations of *Moina mongolica* (after Daday, 1901; illustrated by W. Vars). *A.* Female. *B.* Female antennule. *C.* Head of male. *D.* Male. *E.* Female postabdomen. *F.* Female postabdomen. *G.* Male postabdomen. *H.* Male first leg.

deutlichen Winkel. Bauchrand schwach gewölbt, fast gerade, kahl und erreicht den Hinterrand in einem stumpf abgerundeten Winkel.

Abdomen von gewöhnlicher Form; die Oberfläche seines grösseren und breiteren proximalen Theiles ist mit Gruppen von kleinen Dornen bedeckt; der kleinere und schmälere distale Theil ist seitlich ausser dem Gabeldorn mit 8–10 einfachen, gefiederten Doren bewehrt und am inneren Rande mit einer Querreihe kleiner Dörnchen besetzt. An der schwach gebogenen Endkralle stehen feine Borstenreihen, an der ziemlich kurzen Basis hingegen Borstenbüschel.

Länge: 0.8–1.2 mm; Kopflänge: 0.35 mm; Kopfhöhe; 0.42 mm.

♂. Körper im Ganzen kleiner und kürzer als beim Weibchen. Der Kopf ist vom Rumpfe nur wenig abgesondert, weil ihre Contouren in einer Linie liegen und zwischen beiden nur eine sehr geringe Ausbuchtung besteht. Kopf ziemlich lang, viel länger als die halbe Körperlänge, nahezu kegelförmig, oben bis zur Augengegend schwach abschüssig und gerade, in der Nähe des Auges stärker vorspringend, spitzig und gleichmässig abgerundet.

Die 1. Antennen sehr lang, nahezu von halber Körperlänge, sichelförmig einwärts gebogen und an der Spitze mit Hakendornen bewehrt. Das Auge liegt auf der verengten Stirn, ist aber von dessen Wandung entfernt gelegen.

Der 1. Fuss hat eine eigenthümliche Structur, da an seinem letzten Gliede keine Borste länger ist, als die starke sichelförmige und glatte Endkralle.

Oberfläche der Schale glatt, Rücken—und Bauchrand gerade, fast parallel; beide stossen mit dem hohen, verticalen Hinterrande nahezu unter rechtem Winkel zusammen. Bauchrand fein beborstet.

Das Abdomen ist dem des Weibchens ganz ähnlich.

Länge: 0.92 mm; Kopflänge: 0.47 mm; Kopfhöhe: 0.38 mm. (Daday, 1901: pp. 444–445).

Female

The head of the female is broadly rounded but has an indication of a supraocular depression. Sars (1903a) found only a slight supraocular depression on his *microphthalma*, but there appears to be a definite depression on specimens from North Africa (Gauthier, 1954). The eye is rather small and lies close to the anterior margin of the head. The antennules are short and thick. There is a row of long hairs on the posterior margin of the antennules, and the sensory seta is on the anterior margin at a point almost two-thirds the distance from the head. There are also rings of short setae on the antennules.

The second antennae are long and well developed. The three sensory setae on the basipod are long although the two basal setae are not as long as the basipod segment. There is an inner row of hairs on the two rami, and the exopod ramus lacks the inner row of teeth. The swimming setae are of the usual length.

The shell is broadly rounded especially when the broad pouch is filled with embryos. There are no hairs on the surface of the shell, and the shell is only faintly reticulated. The ventral shell rim has twenty-nine or thirty long setae. The posterior rim has a row of short setae that may either be grouped or ungrouped. There are no shell hooks at the dorsal end of the posterior margin.

The first trunk limb lacks the anterior seta on the third or penultimate segment (fig. 30, *A*). The anterior seta of the terminal segment is only one-half the length of the two feathered setae. However, the ejector hooks on the basipod are long and well developed. The fifth leg is of the usual form.

The postabdomen lacks long hairs on the dorsal margin but has several horizontal rows of short setae (fig. 30, *A* and *B*). The distal conical part has seven to ten short lateral feathered teeth. There is also a short bident tooth that is placed dorsal to the plane of the feathered teeth. The bident is much shorter than found in any other species of *Moina*, and the two arms are of equal length. The claw has only a small pecten of fine hairs at its base. In some specimens this pecten may be missing.

The ephippium of the sexual female contains one egg and is reticulated over its entire surface with a polygonal pattern. It is usually brown.

The females vary considerably in size. Daday (1901) lists females as small as .8 mm. long, but I have seen no specimens that are as small as this. Gauthier (1954) reports a specimen that is 1.80 mm. long. The females of *mongolica* that I have measured range from 1.16 to 1.55 mm. for parthenogenetic females and 1.06 to 1.08 mm. for sexual females. Sars lists *micropthalma* only as "scarcely exceeding 1 mm." (Sars, 1903a: p. 179).

The Aral Sea specimens were 1.12 to 1.28 mm. long.

Male

The males of *mongolica* are very much like those of *brachiata* and *micrura*. The head is rounded in front but has a distinct supraocular depression above the eye (fig. 30, *D*). The eye is rather large and fills the anterior end of the head. The antennules originate below and slightly behind the eye. There is a bend in the antennule at a point about one-third the distance from the head (fig. 30, *C*). The antennule is longer, relative to body length, than usually found in other species of *Moina*. The short, thick-based sensory seta originates near the knee of the bend on the medial margin. The second seta originates below the first and near the anterior margin. The terminal end of the antennule of Daday's specimens has four hooks. The two males that I have examined from the Aral Sea had five hooks.

The second antennae are like those of the female.

The shell is oblong in form and lacks hairs on the surface. There are about thirty setae along the ventral shell rim. There are no hooks at the dorsal end of the posterior rim.

The first leg is very well developed and has a large, recurved hook on the third segment. The terminal

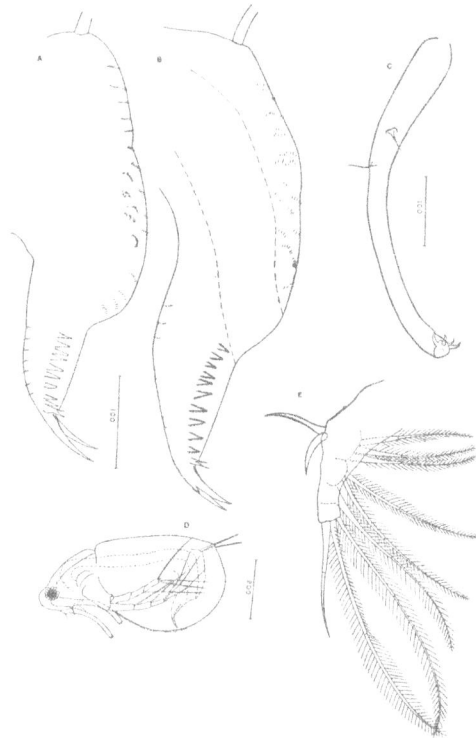

Fig. 30. *Moina mongolica* Daday. *A*. Female postabdomen from Daday's type material, slide number 33-1182. *B*. Female postabdomen, Aral Sea, U.S.S.R., 22-VIII-54 (collected by F. D. Mordukhai-Boltovskoi). *C*. Male antennule, Aral Sea, U.S.S.R., 22-VIII-54 (collected by F. D. Mordukhai-Boltovskoi). *D*. Male, same as *C*. *E*. Female first leg of *A*.

segment is very small and has a small curved seta and two short feathered setae.

The testes lie lateral to the intestine. I could not see the genital opening on the specimens that I examined. The spermatozoa are small spherical cells.

The postabdomen is similar to the female postabdomen.

Two males in Daday's collections measured .91 to 1.0 mm. long. Daday (1901) gives a measurement of .92 mm. According to Gauthier (1954) the males of *salinarum* vary from .83 to .89 mm. long. Sars' male from Egypt measured 1.1 mm. long.

The type material of *M. mongolica* is stored in the Termezettudomany Museum in Budapest, Hungary. Additional material from the Aral Sea has been placed in the British Museum (Natural History).

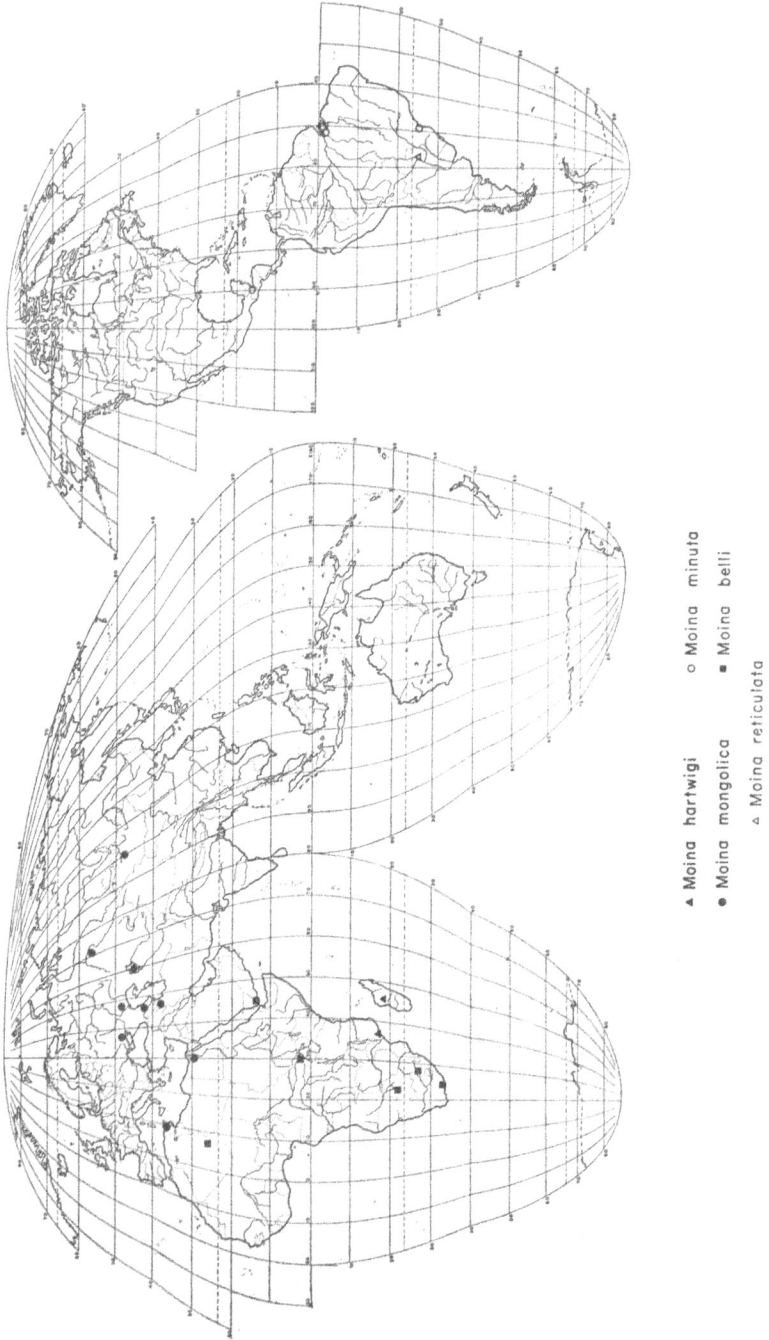

FIG. 31. Geographical distribution of *Moina hartwigi, minuta, mongolica, belli, and reticulata.*

▲ Moina hartwigi   ○ Moina minuta

● Moina mongolica   ■ Moina belli

△ Moina reticulata

DIFFERENTIAL DIAGNOSIS

Because of its large size and restriction to the very saline pools of the Old World, *mongolica* should not be confused with other species of the genus. Occasionally, however, *Moina brachiata* or *micrura* are also found in saline pools. These latter two species can easily be distinguished from *mongolica* by the morphology of the postabdomen. *Moina mongolica* lacks the large claw pecten so characteristic of *brachiata* and has a very short bident tooth while both *brachiata* and *micrura* have a long bident tooth. Furthermore, the females of *mongolica* lack the anterior seta on the penultimate segment of the first foot. It is the only species of *Moina* in the Old World that does. *Moina minuta* also lacks this seta but is found only in the New World. The latter species is much smaller than *mongolica* and has a very different postabdomen.

DISTRIBUTION

Type locality: Chirman-tzagen-nor (Terhin Tsagen Nür), Mongolia.

Moina mongolica appears to be the only halophilic species of *Moina* in the Old World. As such, it is distributed from North Africa, across the Middle East and the Central U.S.S.R. and to Mongolia (fig. 31). Whether or not it is also found in Europe is not presently known. It appears to be particularly common in the Caucasus Region of Central Russia (Behning, 1941). However, some of the reports listed by Behning may be misidentifications of *micrura* or *brachiata*. It is found commonly in the plankton of the Aral Sea (Fadeev, 1925; Zenkevitch, 1963).

Specimens Examined

Mongolia:
Chirman-tzagen-nor, Gobi Desert, 1898 (Daday Collection, slide No. 33–1182).

U.S.S.R.:
Aral Sea, Collection No. 24, depth 60 m., 22-VIII-54
Aral Sea, Collection No. 34, depth 49 m., 22-VIII-54
Aral Sea, Collection No. 23, depth 45 m., 22-IX-57
collected by F. D. Mordukhai-Boltovskoi

Egypt:
Birket Qârûn Fayum (Sars Collection—no date or number).

MOINA BELLI GURNEY, 1904

Moina belli Gurney, 1904: p. 299; pl. 18, figs. 3–4.
Moina macrocopa Sars, 1916: p. 320; pl. 35, figs. 1, 1a.
Moina belli Daday, 1928: pp. 92–93; pl. 5, figs. 1–7.
Moina belli var. *salina* Daday, 1928: pp. 93–94; pl. 5, figs. 8–13.
Moina turkomanica [?] Keiser, 1931: pp. 366–370; figs. 14–21.
Moina turkomanica Jenkin, 1934: p. 160; fig. 12c.
Moina tonsurata Brehm, 1935: pp. 148–150; figs. 3–6.

Moina lateralis Brehm, 1958 [?]; pp. 78–83; figs. 1–6.
Moina sp., *cfr wierzejskii* Harding, 1957; pp. 66–67: figs. 7–9.

DIAGNOSIS

A large species with a rounded head that is almost completely covered with hairs. The antennules are large and robust. The ventral shell rim has from forty to sixty setae followed behind by one or two groups of shorter setae and then by a row of ungrouped setae. The postabdomen has rows of long hairs on the dorsal margin. There are five to seven lateral feathered teeth and one rather long bident tooth on the distal part of the postabdomen. The claw has a very small pecten that consists of several thin teeth.

The ephippium has two eggs and is reticulated with a polygonal pattern.

The mature females range from 1.2 to 1.7 mm. long.

The antennules of the male are bent at a point one-third the distance from the head and have a stout seta originating at this point, on the medial margin. The second seta of the antennule originates at the mid-point of the antennule on the lateral margin. There are five to six terminal hooks on the male antennule. The shell has hairs near the dorsal margin and almost completely covering the shell. The first leg has a well-developed hook and an exopod with a long terminal seta. The male postabdomen has two sperm duct openings near the base of the claw. The spermatozoa are rod-shaped cells as in *macrocopa*.

DESCRIPTION

Gurney's (1904) description of *M. belli* is as follows (fig. 32):

Dorsal margin of head evenly rounded, without any concavity above the eye; ventral margin somewhat protuberant; posterior margin finely ciliated. Fornix well developed and extending over the eye.

Shell without any trace of striation; ventral margin setose for about two-thirds of its length. First antennae ciliated all over. Tail of the usual shape, with eight lateral teeth, the first of which is bifurcated. Between the bifurcated tooth and the first simple tooth is a minute elevation covered with cilia, which may represent a rudimentary tooth. Apical claws armed with a basal row of secondary denticles and with a ventral chitinous expansion cleft into teeth. Posterior dorsal surface of tail provided with cilia, which are more or less arranged in transverse rows. Ephippium reticulated all over and containing two resting eggs.

Length 1.7 mm.

Several specimens of this species were contained in the collection, but all were females, one of which was ephippial. The species very much resembles *M. wierzejskii* Richard, and perhaps should be regarded as only a variety of that species. It is mainly distinguishable by the ciliation of the head and first antennae, and by the structure of the postabdomen. (Gurney, 1904: p. 299.)

Females

The head is rounded; the supraocular depression is almost totally lacking although a slight indentation

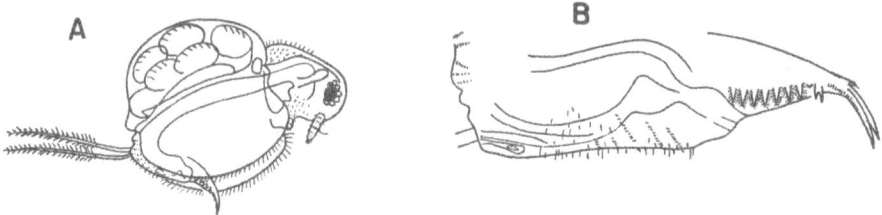

FIG. 32.   Gurney's illustrations of *Moina belli* (from Gurney, 1904).   *A*. Parthenogenetic female.   *B*. Female postabdomen.

may occasionally be seen (fig. 33, *A*). The eye is rather large and lies near the middle of the head. The eye is composed of a very large pigment spot with many small crystalline bodies surrounding it. The antennules are long, very robust, and are frequently folded back against the body. They are ornamented with many rings of short setae. The posterior and medial margins of the antennules are covered with long hairs. The sensory seta originates from the mid-point along the anterior margin.

The second antennae are very well developed, but the individual segments of the rami are rather short. Yet the over-all length of the antenna is comparable to the second antenna length in *macrocopa* and other species.

The two sensory setae on the basipod are short. The individual segments of the rami are covered with many short setae that are arranged in rows around each segment. The medial margin of both rami have many long hairs.

The four-segmented exopod is armed with teeth, which may be in groups of three or four, but extend only over the proximal two-thirds of the segment—the distal third is bare and is often expanded medially. The distal portion of each expod segment is therefore much broader than the proximal part.

The head is almost completely covered with hairs, but these are absent from the anterior margin. The shell lacks hairs on its posterior half. The hairs are particularly dense on the back of the head and on the antero-ventral part of the valves. Occasionally these hairs may be sparse as in Harding's (1957) specimen from Albert National Park in Central Africa.

The shell is reticulated as usual—not lacking any trace of striation as stated by Gurney (1904). The shell does not have any other markings or gland cells as in *wierzejskii* or the other species.

The ventral shell rim is completely lined with forty to sixty long setae. The setae are longer near the anterior margin and become short and closely spaced near the posterior margin. The posterior shell rim has one or two groups of unequal-sized setae followed by a row of many equal-sized setae. Sometimes the groups of setae are quite evident, and the largest setae of each group appear as teeth projecting from the

shell. This is particularly true of specimens from Aden. However, this is a variable character in the species.

There is a pair of elbow-shaped shell hooks near the dorsal end of the posterior margin.

The first leg is not like the first leg of *macrocopa* but instead has the normal setation found in *australiensis* as well as most other species of *Moina* (fig. 33, *B*). The anterior seta of the penultimate segment has only fine hairs along its length as does also the anterior seta of the ultimate segment.

The postabdomen is very long and robust (fig. 33, *C*). The distal conical portion composes only a small part of the total length of the postabdomen. The dorsal side is ornamented with rows of hairs that extend from one side to the other. The long hairs are found only on the proximal half of the postabdomen— these are replaced by short setae towards the distal end. The conical part of the postabdomen carries only five to eight lateral feathered teeth and one bident tooth that is placed slightly dorsal to the plane of the other teeth. No other large species of *Moina* has so few feathered teeth. Frequently there will be a small cluster of setae between the bident tooth and the distal feathered tooth. This cluster may appear at times to be an additional tooth (see remarks by Gurney, 1904).

The claw has a "Basaldorn" of four or five teeth and a pecten of many very fine teeth as in *macrocopa*. The distal part of the claw has a row of fine setae or may have a row of stout teeth. The latter character may be seen on specimens from Aden.

The ephippium of the sexual female is reticulated with a polygonal pattern that is either restricted to the area over the two ampulae or may be restricted to the marginal areas. The first condition exists on specimens from South Africa, the second on specimens from Aden.

The parthenogenetic females vary in length from about 1.2 to 1.7 mm. The sexual females are usually smaller, rarely exceeding 1.3 mm. in length.

Male

The male of *Moina belli* is intermediate in form between the males of *australiensis*, *tenuicornis*, and *macrocopa.* The antennules originate slightly behind the anterior tip of the head, and the large eye fills the end of the head (fig. 34, *A*). There are few hairs on the head.

The antennules are bent at a point one-third the length of the antennule from the head (fig. 34, *C*). In *tenuicornis* this bend is nearer the head while in *macrocopa* it is at the mid-point of the antennule. The antennules are broadly curved distal to this bend to form the clasping organs for holding the female. The distal end of the antennule has five or six short hooks.

There are two sensory setae on the antennules, as in all other species of the Family Moinidae. One seta originates at the knee of the bend, but the second is located along the lateral side near the mid-point of the antennule (fig. 34, *C*). This would be comparable to its point of origin on the male antennule of *macrocopa.*

The posterior margin of the antennule has several rows of long hairs along its length. The second

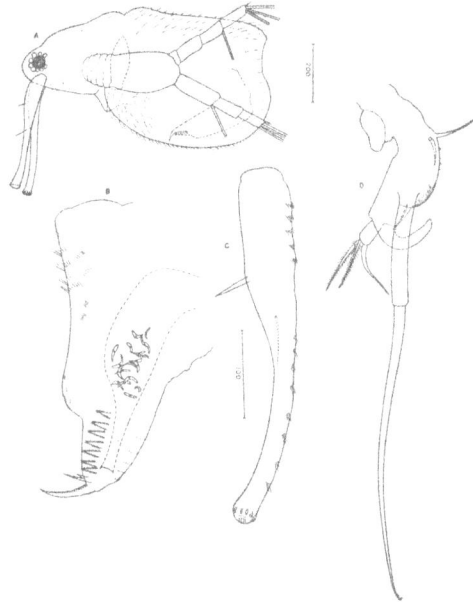

FIG. 34. *Moina belli* Gurney. *A*. Male from Kalahari Desert, Bechuanaland (Daday Collection). *B*. Postabdomen of *A*. *C*. Antennule of *A*. *D*. First leg of *A*.

antennae of the male are much like those of the female although the distal ends of the exopod segments are not as expanded as in the female.

The shell is oblong and has approximately fifty setae on its ventral margin. The setae on the posterior margin are usually ungrouped and are not as densely packed as in the female.

The shell is almost completely covered with hairs. These hairs are particularly dense near the anterior and ventral margins.

The first leg of the male is very similar to the first leg of *macrocopa* (fig. 34, *A*). The hook is large although not as stout as the hook of *macrocopa*. The terminal seta on the exopod is very long and projects beyond the posterior margin of the shell.

The male postabdomen is very similar to that of *macrocopa* (fig. 34, *B*). The vas deferens opens on either side, dorsal to the base of the claw. The spermatozoa may be seen in the testes which lie on either side of the gut. The spermatozoa are sickle or rod shaped as in *macrocopa.*

The type material of this species is in the British Museum (Natural History).

DIFFERENTIAL DIAGNOSIS

Moina belli sould not be confused with many other species because it is restricted to Africa and part of the

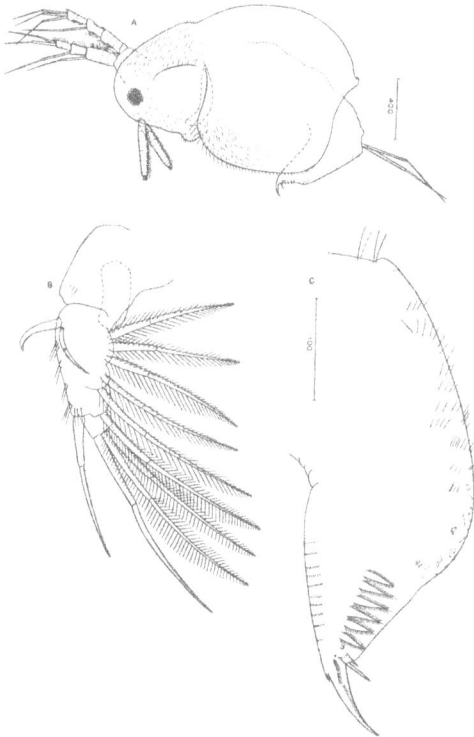

FIG. 33. *Moina belli* Gurney. *A*. Parthenogenetic female from Kalahari Desert, Bechuanaland (Daday Collection—no locality or date given). *B*. Female first leg, water tank, Aden, 7-XII-32 (collected by G. Evelyn Hutchinson). *C*. Postabdomen of *B*.

Middle East area. It shows a very great resemblance to *macrocopa* and is undoubtedly closely related to this species. However, *belli* lacks the teeth on the two anterior setae of the penultimate and ultimate segments of the first leg that are so characteristic of the female of *macrocopa*. In addition, the postabdomen of *belli* has only five to seven lateral feathered teeth while *macrocopa* generally has seven to ten. However, a few specimens of *belli* from South Africa have up to eight lateral teeth on the postabdomen.

The ephippium of *belli* has a slightly different pattern from that of either *macrocopa* or *tenuicornis* from the Old World. The ephippium is ornamented with a polygonal pattern; in the other two species this pattern is of rectangular reticulations that are in rows somewhat like bricks. *Moina macrocopa americana* from the New World, however, has an ephippium similar to that of *belli*.

The male antennule of *belli* is bent at a point one-third the distance from the head, and only one seta originates at this bend. The second seta originates at about the mid-point in length of the antennule. In *tenuicornis* the bend is near the head, and the two setae are close together at this point. In *macrocopa* the antennules are bent at the mid-point, and again both setae originate here.

The males of *macrocopa* and *belli* differ little in other characters although the head of *belli* is not covered with hairs as it may be in *macrocopa*.

The male postabdomen is very similar to the postabdomen of *macrocopa* and therefore differs from *tenuicornis* in that the openings of the vas deferens are very pronounced and located near the base of the claw. The spermatozoa are sickle shaped as those of *macrocopa*; whereas the spermatozoa of *tenuicornis* are small spherical cells.

The males and females of *Moina belli* differ in only a few characters from *macrocopa*, but it is quite apparent that the two are distinct species. The variant characters in *belli* approach characters found in *tenuicornis*. The immediate impression is that *belli* is an intermediate species that may have originated from a *tenuicornis* type species, and itself may have given rise to *macrocopa*. The suggestion that both *belli* and *macrocopa* are somewhat more "advanced" than *tenuicornis* or some of the other species is suggested by their spermatozoa which are quite unique for the genus.

DISTRIBUTION

Type locality: A water-hole on the veld at Kroonstad, South Africa. It is impossible to give, with much accuracy, the geographical distribution of *Moina belli* (fig. 31). In addition to the type material, I have examined specimens only from the Kalahari Desert in South Africa, the Congo Republic, and Aden. It appears as if Brehm's species *lateralis* from Algeria

is *belli*, and it is possible that Keiser's (1931) *turkomanica* from the Kara Kum Desert of Southern Russia may also be *belli*. However, this cannot be verified until Keiser's type material is found or material from this region is available. There are reports of *belli* from Greece by Stephanides (1936, 1937, 1948) that have not been confirmed although I am inclined to believe his form to be *macrocopa* (his figures of the male antennules are much more like *macrocopa*).

If most of these latter records prove to be of *belli*, then it would seem probable that this species is widespread throughout the Middle East and Southern Russia. We do know that *macrocopa* is found from North Africa, various parts of the U.S.S.R., Europe, India, China, and Japan.

The geographical ranges of these two species may overlap to some degree although *belli* may prove to be restricted to the desert and arid regions. Most reports of *belli* are, in fact, from such regions.

Specimens Examined

Africa:

Bechuanaland:

Kalahari Desert (Daday Collection, no date or locality given)

Port Elizabeth (Sars Collection, no date or locality given)

Water hole on the veld at Kroonstad (type material of Gurney, in British Museum, Natural History)

Congo Republic:

Temporary pond in Albert National Park, 8-VIII-47 (British Museum, Natural History, collection 1956.7.4.14)

Aden:

Water tanks near boat docks, 7-XII-32 (collected by G. Evelyn Hutchinson)

MOINA RETICULATA (DADAY, 1905)

Moinodaphnia reticulata Daday, 1905: pp. 202–203; pl. 13, figs. 6–8.

DIAGNOSIS

The body is round, quite similar to *Ceriodaphnia*, but there is a slight keel along the dorsal margin of the shell. The head is small, with a large compound eye and an ocellus. The antennules and head are bent ventrally.

The second antennae are of the usual form but the exopod ramus lacks a medial row of stout setae.

The abdomen has a large horseshoe-shaped fold on the dorsal half that serves to hold the embryos in the brood pouch. The shell lacks hooks at the dorso-posterior angle.

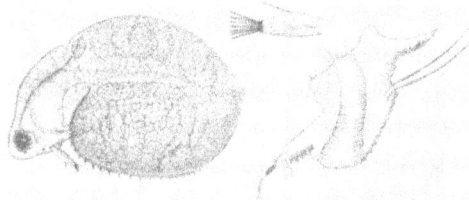

FIG. 35. Daday's illustrations of *Moina reticulata* (from Daday, 1905). *A*. Female. *B*. Postabdomen.

The postabdomen has five or six lateral feathered teeth and a single bident tooth that has two spines of equal length.

The females are .6 to .7 mm. long.

Sexual females and males are unknown.

DESCRIPTION

The following is Daday's description of *M. reticulata* (fig. 35):

Der Körper ist im ganzen annähernd eiförmig, hinten etwas breiter; zwischen Kopf und Rumpf scharf einge-schnitten.   Der Kopf ist nach vorn und etwas nach unten gerichtet, ober dem Auge zeigt sich eine scwache Vertie-fung, demzufolge die eigentliche, das Auge in sich schlies-sende Stirn ziemlich abgesondert ist; ober der Vertiefung ist der Kopfrand schwach bogig.   Die Stirn ist nach vorn gerichtet und ziemlich spitz gerundet.   Der Bauchrand des Kopfes ist am Ausgangspunkt der ersten Antennen etwas ausgebuckelt.   Der Fornix entspringt ober dem Auge, zieht anfänglich parallel des Kopfrandes hin und bildet dann unter der Vertiefung einen Lappen.

Das Auge besteht aus vielen Linsen, ist gross und füllt die Stirnhöhle fast ganz aus.   Der Pigmentfleck ist sehr klein, fast viereckig und liegt nahe der Basis der ersten Antennen.   Die ersten Antennen sind im Verhältnis kurz, überragen die halbe Länge des Bauchrandes des Kopfes nicht, sind annähernd spindelförmig mit glatter Ober-fläche, an der Basis der Riechstäbchen erhebt sich ein Kranz kleiner Dornen.   An den Astgliedern des zweiten Antennenpaares zeigen sich 2–3 kleine Dornkränze.

Der Rückenrand der Schale ist schwach bogig, gegen den Hinterrand abschüssig und mit demselben eine kleine Spitze bildend.   Der Hinterrand ist sehr kurz und übergeht unbemerkt in den schwach gerundeten Bauchrand, welcher vom hinteren Drittel an nach vorn schwach abschüssig und in der ganzen länge mit spärlich stehenden kleinen Haaren besetzt ist.

Die Schalenoberfläche ist am Rumpf mit regelmässigen sechseckigen Felderchen geziert, deren Innenraum fein granuliert erscheint.

Der Abdominalfortsatz ist dick, fingerförmig.   Das Postabdomen ist kegelförmig und in eine breite supraanale und eine schmale anale Partie geteilt; am Hinter- bezw. Oberrand des supraanalen Teiles zeigen sich gleich weit voneinander stehende punktförmige Kutikularverdickun-gen und bildet an der Grenze des analen Teiles einen stumpf gerudeten Winkel.   An beiden Seiten des analen Teiles stehen ober dem distalen Gabeldorn sechs einfache, borstenartige Dornen.   Die Endkralle ist sichelförmig, die Aussenseite trägt in der proximalen Hälfte einen Kamm von kräftigeren Borsten und ist fernerhin fein behaart. An der Basis der Endkralle erhebt sich vorn ein Dorn.

Die Körperlänge beträgt 0,65–0,7 mm, die grösste Höhe 0,24 mm.   (Daday, 1905: pp. 202–203.)

Female

The body is rounded and has the head depressed (fig. 36, *A*).   The antennules are frequently folded backward and are not always readily visible.   The surface of the shell and head is reticulated, but there are no hairs.   The compound eye is large and fills the tip of the head.   According to Daday (1905) an ocellus is present.   This could not be seen on his type mate-rial, but it is not unusual for an ocellus to be lost when animals are killed and preserved.

The antennules are short in contrast to the other species of *Moina*.   The sensory seta originates from the front margin, and there is a bundle of long sensory papillae on the distal tip.

The second antennae are not well developed; the swimming setae reach only to the middle of the shell. The sensory setae on the basipod are only moderately long.   The swimming setae pattern is like that of the other moinid species.   The distal spine on the exopod has not increased in length as is typical of *Moino-daphnia*.   The exopod lacks the medial row of teeth-like setae.

The shell is without surficial hairs.   There is a keel along the dorsal margin that makes the animal flat-tened laterally although this development is not so extreme as in *Moinodaphnia macleayi*.

The ventral margin of the shell has a row of about fifteen setae.   Behind this row, the margin is armed with fine setae; but these are not grouped together as in *Moina micrura*.   There are no hooks at the dorsal end of the posterior shell margin.

The first leg has the usual setation pattern.

The postabdomen has a short pair of setae at the proximal end and is ornamented with fine hairs along its dorsal margin but lacks long hairs (fig. 36, *B*).

The distal half of the postabdomen has a row of five to six feathered setae and a long bident tooth that has both spines equally long.   The proximal end of the claw has a short pecten of ten to twelve setae and the distal part of the claw is armed with a row of thin setae.

The posterior margin of the brood pouch is closed off by a large horseshoe-shaped fold of the abdominal epidermis.   This fold extends from the lateral sides of the abdomen back to the distal end of the abdomen where it extends to the dorsal surface.   The dorsal surface of the abdomen also forms the placenta or "Nährboden" characteristic of the Moinidae.

The females are .6 to .7 mm. long.

Males and sexual females have not been collected as yet.   The type material for this species is stored in the Termeszettudomany Museum, Budapest, Hungary.

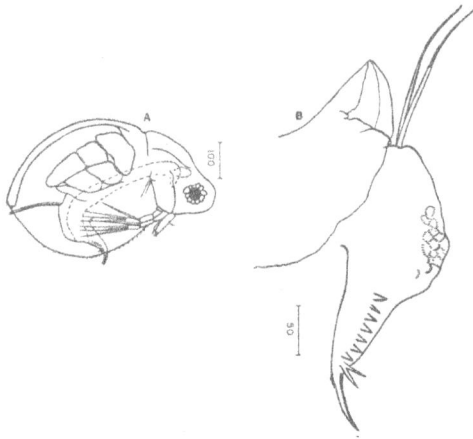

FIG. 36. *Moina reticulata* Daday. *A*. Parthenogenetic female, Paraguay (Daday Collections, no locality or date given). *B*. Postabdomen of *A*.

Moina reticulata has only been reported by Daday from several inundation pools in Paraguay. Daday's type material was used in the above description.

Daday placed this species in the genus *Moinodaphnia* because it possessed the ocellus and a well-developed abdominal fold. It does not, however, have the long spine on the exopod ramus of the second antenna which is a very important character of *Moinodaphnia*.

In the present work this species has been placed in the genus *Moina* for two reasons. First, the presence of an ocellus and of the abdominal fold are not infallible characters since it is possible that one species of *Moina*, *minuta*, has an ocellus while several species, most notably *weismanni*, have the abdominal fold. Secondly, the species lacks the characteristic pattern of setae on the shell and second antennae. Furthermore, the form of the shell and head are quite different in *Moina reticulata* from what is observed in *Moinodaphnia*.

Before *reticulata* can be placed in the genus *Moinodaphnia* it should be found to have similar males and sexual females, and it should have a similar habitat as *macleayi*. The latter is a littoral form which frequents weeds and commonly attaches itself to plants where it then filters the water surrounding the plant (Sars, 1901). It is not, therefore, a member of the zooplankton as the species of *Moina* are.

DIFFERENTIAL DIAGNOSIS

Moina reticulata may be distinguished from *M. micrura* by the presence of an ocellus, the well-developed abdominal fold, and the long bident tooth on the postabdomen which has both spines of equal

length. The same features serve to distinguish the species from most other forms of *Moina*.

Moinodaphnia macleayi is a larger species and has a more distinct keel on the dorsal margin of the shell. Furthermore, the head of *M. macleayi* is not depressed or round but is a triangular shape. The postabdominal claw of *macleayi* lacks a pecten.

Moina reticulata is a distinct species and should not be confused with other species of the family.

Its precise position and relationship within the family is not known and will not be known until the secondary sex characters of sexual females and males are described.

DISTRIBUTION

Daday listed the following localities in Paraguay for this species:

Asuncion, Campo Grande, Calle de la Cañada, von Quellen gebildete Tümpel und Gräben; Curuzu-chica, toter Arm des Paraguayflusses; Curuzu-nú, Teich beim Hause des Marcos Romeros; Lugua, Pfütze bei der Eisenbahnstation; Paso Barreto, Bañado und Lagune am Ufer des Rio Aquidaban; Villa Sana, Inudationen des Baches Paso Ita und Pegauho-Teich. (Daday, 1905: p. 203.)

It has not been reported subsequent to Daday's publication (fig. 31).

Specimens Examined

Paraguay:

No locality given (Daday Collection No. 1917-37).

MOINA HUTCHINSONI BREHM, 1937

Moina hutchinsoni Brehm, 1937a: pp. 91–94; figs. 1–8.
Moina hutchinsoni Brooks, 1959: pp. 622; fig. 27.47.

DIAGNOSIS

The head is rounded but occasionally with a slight supraocular depression. The eye is of moderate size. The antennules are short and thick. The second antennae are well developed. The shell has hairs on its surface near the antero-ventral margin. The ventral shell rim is completely lined with about forty setae that extend onto the posterior margin. Behind these setae there is a row of short setae, all of equal size. The shell hooks are absent. The distal tooth on the postabdomen is unident rather then bident or is completely absent.

The males have long antennules that are bent at a point very near the head. The tip of the antennule has five hooks. The first leg has a small distal hook. The two seminal openings are near the base of the claw. The spermatozoa are large spherical cells.

DESCRIPTION

This species was first found in collections from Soda and Winnemucca lakes in Nevada that had been sent to Brehm by Professor G. Evelyn Hutchinson—hence the specific name. Brehm described the animals from the lakes as two distinct races. No males were present in these collections. I have only seen the specimens from Winnemucca. However, Professor W. T. Edmondson has kindly sent me both preserved and live material of *Moina hutchinsoni* from Soap Lake, Washington. These collections contained many male specimens that have been used in the present description. The Soap Lake specimens are a distinct race from both the Soda and Winnemucca forms. Therefore, greater variability should be found in male specimens than is indicated below. I shall refer, in the present description, to each race by the name of the lake from which it was collected. However, these should probably be considered as distinct subspecies.

Female

A very large and robust species (fig. 37, *A*). The head is large but not oversized in comparison to the body. There may be a slight indentation above the eye. According to Brehm (1937), the race from Winnemucca has a slight supraocular depression but the race from Soda Lake does not. The eye also varies in size—it is of a moderate size in the Soda and Soap lake races but is small in the Winnemucca race.

The head may be completely covered with hairs (Soap Lake) or only partly covered with hairs (Winnemucca).

The labial keel is very large and has hairs on its ventral margin.

The antennules are short, somewhat stubby, and have rings of setae along their full length (fig. 37, *D*). There is a vertical row of long hairs on the posterior margin of the antennule. The sensory seta of the antennule is long, about two-thirds the length of the antennule. The sensory papillae on the tip are thick, and there appear to be only five or six papillae rather than the usual nine.

The second antennae are very pubescent. The three sensory setae on the basipod, two at the base— one at the distal margin, are all long. The medial margin of the exopod ramus has a row of teeth that extends the full length of each segment. The swimming setae are of the usual length in proportion to the length of the segments.

The shell is covered with hairs in the Soap Lake race but only a few are present on the shells of the Winnemucca race. These hairs may be grouped in horizontal rows near the anterior and ventral margins and are very long. Behind this area the hairs are short and are dispersed.

The ventral shell rim is lined with long setae that extend from the anterior margin onto the posterior margin. There are from thirty-six to forty setae. Behind these setae there is a row of shorter setae that are spaced relatively far apart.

The shell lacks the pair of hooks at the dorsal end of the posterior shell margin.

The first leg is well developed and has the anterior setae on the third and terminal segments. These setae are feathered with rows of short hairs. The ejector hooks are long and are also well developed.

The postabdomen is very large and pubescent (fig. 37, *B* and *C*). It is covered behind with many rows of short setae that interconnect to present the usual pattern of short setae. On the Soap Lake specimens, there may be a few long hairs near the base of the long abdominal setae. Otherwise, these hairs are absent.

The distal part of the postabdomen has a varied armature. The Soda Lake specimens have a row of seven to nine lateral feathered teeth and a single distal unident tooth; the latter represents the bident tooth. The Winnemucca forms also have the unident tooth and have nine to ten feathered teeth. The Soap Lake forms have only five to eight feathered teeth and completely lack the unident tooth. In this regard they are like *Moina eugeniae*. However, one specimen from Soap Lake was found to possess a greatly reduced bident tooth (fig. 37, *E*). This type of aberration is common in the genus.

The claws have a pecten of several thin hairs or setae that are grouped at the base of the claw. There is a row of fine hairs distal to this pecten. The ventral side of the claw has a "Basaldorn" of four or five teeth.

The ephippia of the sexual female are not as well marked as in other species of the genus (except *eugeniae*) (fig. 37, *A*). Rather, the shell chitin is only slightly thickened and is ornamented with knoblike protuberances that are spaced much farther apart than on ephippia of *M. brachycephala* or *M. wierzejskii*. There are two eggs in the ephippium—*eugeniae* has one.

The parthenogenetic females measure 1.4 to 1.6 mm. long.

Males

The head lacks a supraocular depression, and the eye does not quite fill the end of the head. The antennules are long and broadly curved (fig. 38, *A*). There is a slight bend near the head where the broad based sensory seta originates. The second sensory seta is on the lateral side behind the other seta. The medial margin of the antennules has several groups of short hairs. The terminal end of the antennule has five short recurved hooks.

The second antennae are similar to those of the female.

The shell of the male is covered with hairs. These are particularly dense near the anterior and ventral margins where they are longer and are arranged in rows.

The ventral shell rim is armed with setae as in the female, and there are no hooks present at the dorsal end of the posterior margin.

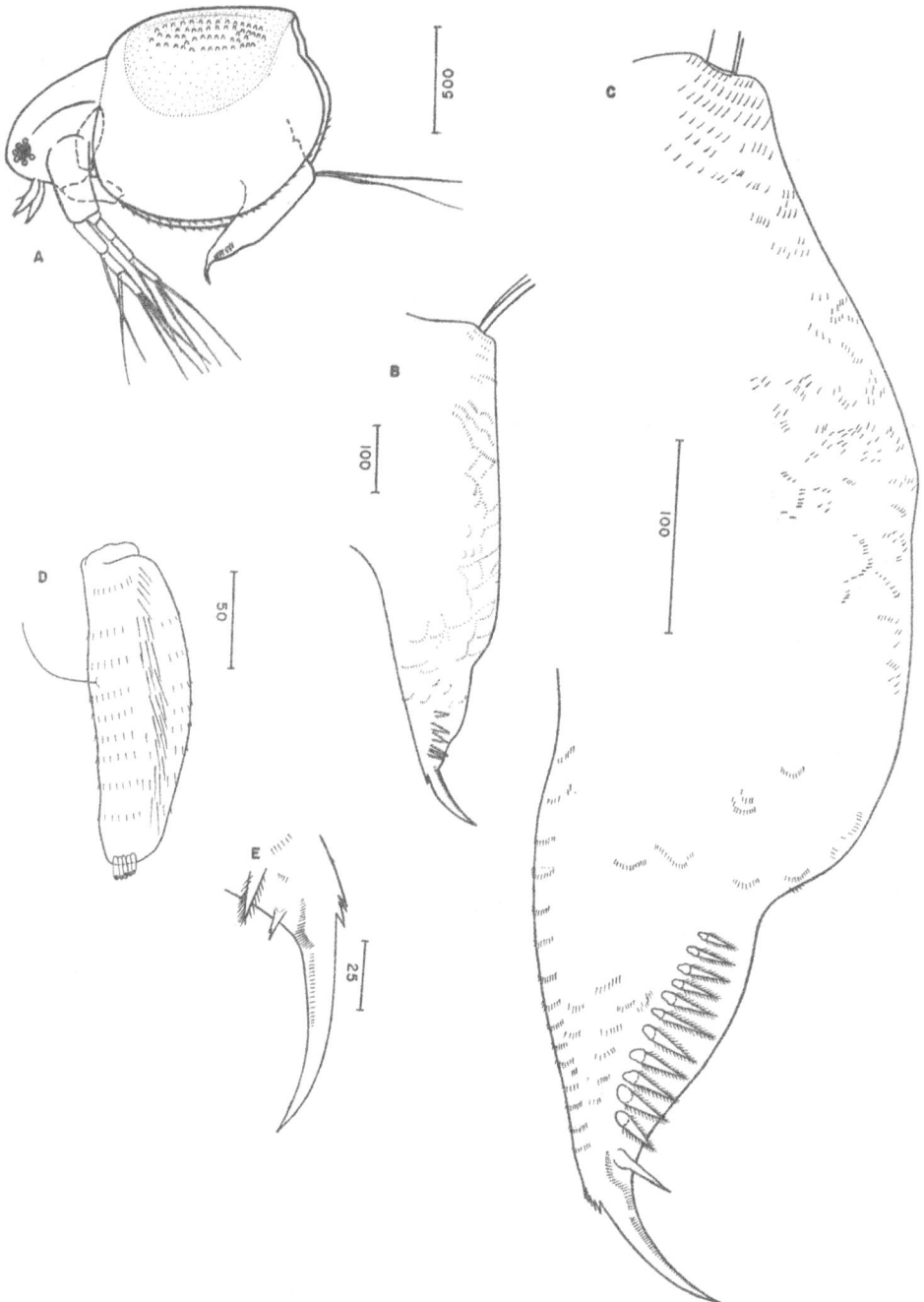

Fig. 37. *Moina hutchinsoni* Brehm. *A.* Sexual female from Soap Lake, Washington, 18-VII-63 (collected by W. T. Edmondson). *B.* Postabdomen of female from Lake Winnemucca, Nevada (collected by G. Evelyn Hutchinson; type material of Winnemucca Race of *M. hutchinsoni* described by Brehm, 1937). *C.* Postabdomen of female, Lake Winnemucca, Nevada. *D.* Antennule of C. *E.* Postabdominal claw, Soap Lake, Washington, 18-VII-63.

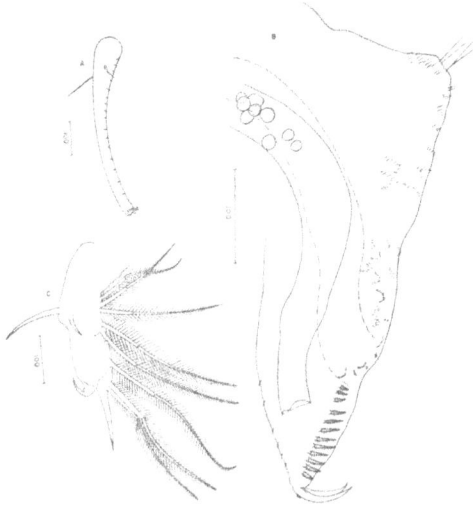

FIG. 38. *Moina hutchinsoni* Brehm. *A*. Male antennule, Soap Lake, Washington, 12-X-62 (collected by W. T. Edmondson). *B*. Postabdomen of *A*. *C*. First leg of male, Soap Lake, Washington 11-XI-61 (collected by W. T. Edmondson).

The first leg has a small distal hook on the third segment (fig. 38, *C*). There is no exopod on this leg. The terminal segment has a long spinelike seta that is slightly curved. The two feathered setae of the terminal segment originate from the side of this spinelike seta. Otherwise, the setation pattern is similar to that normally found on the first legs of *Moina* males.

The postabdomen differs little from that of the female. Male specimens collected from Soap Lake in October had seven to eight teeth while specimens collected in November had as many as twelve teeth. Whether this is a seasonal difference is unknown.

The testes lie lateral to the intestine and extend onto the postabdomen. There are two seminal openings, one on either side of the postabdomen, and located near the base of the claw but not as close to the base as it is in *belli* and *macrocopa* (fig. 38, *B*). The spermatozoa are large spherical cells with no protoplasmic extensions.

Topotype material of this species has been placed in the United States National Museum.

I have included below Brehm's description of the two races in order to delineate further the three races. The races have not been assigned new names but rather are referred to in association with the name of the lake from which they were collected. Further work on these forms would not only be desirable but also very enlightening.

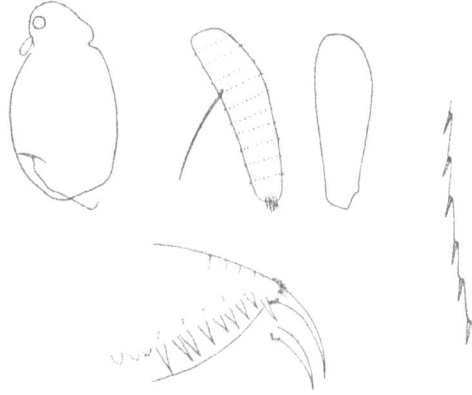

FIG. 39. Brehm's illustrations of *Moina hutchinsoni* from Soda Lake, Nevada (from Brehm, 1937).

Rasse aus dem Soda-Lake [fig. 39].

Die geschlechtsreifen Tiere hatten eine durchschnittliche Grösse von 1600 μ. Nach der von Penelope Jenkin vorgenommenen Klassifikation der Körperformen (vgl. Cladocera from the Rift Valley-Lakes in Kenya in Ann. a. Mag. Nat. Hist. Ser. X, 13. 1934) gehört unsere Form zu den 〉〉depressed Forms〈〈. Das Auge ist verhältnismässig klein. Die Antennulae sind kurz und breit, etwas gekrümmt und tragen die lange Sinnesborste im ersten Drittel. Ferner sind sie mit etwa 10 nicht sonderlich auffallenden Borstenkränzen versehen. Die Antenne trägt am dreigliedrigen Ast und an den beiden basalen Gliedern des viergliedrigen Astes quergestellte Börstchenkämme, während diese an den beiden distalen Gliedern des viergliedrigen Astes zu je einer Längsreihe umgruppiert sind. Der ventrale Schalenrand ist mit kurzen, dräftigen Stacheln bewehrt.

Das Postabdomen besitzt in feine Spitzen auslaufende Endkrallen, an deren Basis ein Kamm stärkerer Stacheln sitzt. Es ist mir zweifelhaft, ob dieser mit dem Nebenkamm, wie ihn viele *Moina*-Arten besitzen, identisch ist; da er nicht der Krümmung der Kralle folgt, sondern einen entgegengesetzt konvexen Bogen bildet, und da er ferner sehr proximal gelegen ist, möchte ich ihn für homolog mit jenem Gebilde halten, das W. Rammner (Zur Variabilität von Daphnia pulex in Internat. Rev. d. Hydrobiol. **29**, 81) als proximalen Nebenkamm bei den Daphnien bezeichnet hat. Die Zahl der bewimperten Zähne an den beiden Seiten des Postabdomens schwankt von 7 bis 9. Was beim Abdomen am meisten auffällt, ist der Umstand, dass der bekannte Gabelzahn, der proximal vor den Wimperzähnen steht, und der als Kennzeichen der Gattungen *Moina* und *Moinodaphnia* gilt, hier ganz atypisch entwickelt ist. Dabei ist auffallend, dass die hier an Stelle des Gabelzahnes auftretenden Ersatzegebilde in ihrer Form ausserordentlich veriabel und bei allen untersuchten Exemplaren auf der rechten und linken Flank des Postabdomens verschieden ausgebildet sind. Als Norm kann etwa gelten, dass auf der einen Seite ein langer einfacher Dorn an Stelle des Gabelzahnes steht, während auf der gegenüberliegenden Seite nur ein kurzer, auf einem Höcker stehender Zahn angetroffen wird. Aber bei einigen Tieren fand sich auf der einen Seite ein kurzer Dorn, während die Gegenseite an dieser Stelle ganz unbewehrt war, bei einem anderen Exemplar entsprach dem kurzen Dorn auf der einen Seite gegenüber eine Gruppe von zwei oder drei kleinen Stacheln. In zwei Fällen zeigt sich als Pendant zu dem kleinen Dorn ein mittellanger Stachel, der am Ende in zwei divergierende Spitzen ausgezogen war, also gewissermassen den Anlauf zu einer Gabelzahnbildung aufwies. Dies Deutung setzt natürlich die Annahme voraus, das hier ein im Entstehen befindliches Merkmal vorläge, eine Annehme, die ebensogut möglich wäre wie die, dass es sich um eine Rückbildungserscheinung handle. Ja, man könnte mit Rücksicht auf die Milieuverhältnisse eher mit der zweiten Möglichkeit rechnen, wofür auch die Inkonstanz der Bildungen spricht, die zum Teil einen fast pathologischen Eindruch machen. (Brehm, 1937: pp. 91–93.)

Moina Hutchinsoni, Rasse aus dem Winnemucca-Lake [fig. 40].

Gegenüber der Rasse aus dem Soda-Lake ist diese Rasse, wie schon erwähnt, dadurch gekennzeichnet, dass an Stelle des Gabelzahnes bei allen Individuen ein einfacher langer Stachel vorhanden ist. Aber abgesehen davon unterscheiden sich die beiden Rassen noch in nachstehenden Punkten, von denen speziell die ganz ungewöhnliche Kleinheit des Auges bei der vorliegenden Rasse auffällt. Leider fehlten auch von dieser die Ephippialweibchen und-männchen. Es ist sehr leicht möglich, dass deren Auffindung zu dem Resultate führen würde, dass die hier als Rassen unterschiedenen Formen zwei ganz verschiedene Arten sind. (Brehm, 1937: pp. 93–94.)

DIFFERENTIAL DIAGNOSIS

Moina hutchinsoni and *eugeniae* are easily identified by the absence of the bident tooth on the postabdomen. These two species may in turn be distinguished by their size, *M. hutchinsoni* usually measures over 1.4 mm. long while *eugeniae* is 1.0 mm. or less. In addition, *M. eugeniae* has a narrow head and a supraocular depression while *hutchinsoni* has a broad head and lacks the depression.

Moina hutchinsoni is a very transparent form that is restricted to very saline and alkaline lakes. No other known species of *Moina* occurs in similar habitats in North America.

DISTRIBUTION

This species is found only in the Western United States and thus far it has been reported only from closed basins in the Great Basin and Grand Coulee Regions of the Far West (fig. 42).

Specimens Examined

North America:
  Nevada:
    Winnemucca, 1930, collected by G. Evelyn Hutchinson
  Washington:
    Soap Lake, 12-X-62, 11-XI-62, 18-VII-63, collected by W. T. Edmondson.

MOINA EUGENIAE OLIVIER, 1954

Moina eugeniae Olivier, 1954: pp. 81–86; figs. 1–8.
Moina eugeniae Olivier, 1962: pp. 214–215; pl. 12, figs. 3–4.

DIAGNOSIS

Head with a supraocular depression. The shell rim carries a row of long setae extending from the anterior to the posterior margins. The postabdomen has five to seven feathered teeth but no bident tooth. The ephippium has one egg and is ornamented with knoblike protuberances.

The male has very long antennules that are bent at a point approximately one-fourth the distance from the head. Two sensory setae are present, one at the bend and the second near the head. The terminal end of the antennules has about five hooks. The form of the male first leg is unknown.

DESCRIPTION

I have not seen specimens of *eugeniae* and have had to depend completely on Olivier's (1954) description. Olivier described the species as follows (fig. 41):

Valvas fuertes, ovoidales, con borde posterior romo, sin escultursa; margen ventral más bien redondeado aunque algo contraido en la región media, armado con finas espínulas en toda su extensión [fig. 41, 4]. El borde dorsal de las valvas varia de acuerdo con que si la hembra posee embriones no; en caso de poseerlos, este borde se eleva considerablemente en forma semicircular y en caso contrario es levemente curvado. Cabeze grande en proporción con las valvas, algo deprimida, depresión supraocular muy pequeña o ausente, parte frontal redondeada, lo mismo que el borde dorsal, margen ventral convexo en el lugar de la inserción de las anténulas y de terminación posterior suave [fig. 41,1]. Anténulas de forma común; llevan un largo flagelo sensitivo más o menos en la parte media y 8–10 cerdas sensitivas terminales [fig. 41,5]. Ojo más bien pequeño con lentes bien marcados y ubicado próximo al extremo fronto-ventral de

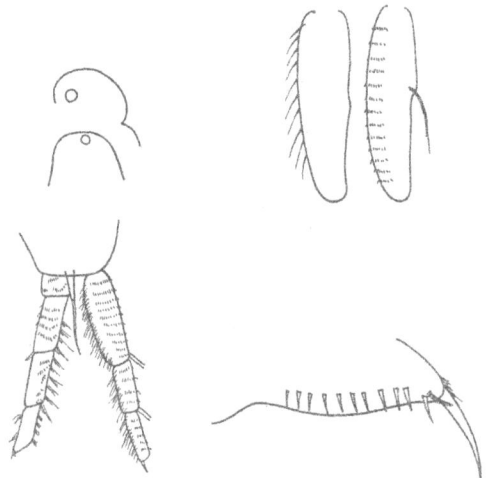

FIG. 40. Brehm's illustrations of *Moina hutchinsoni* from Lake Winnemucca, Nevada (from Brehm, 1937).

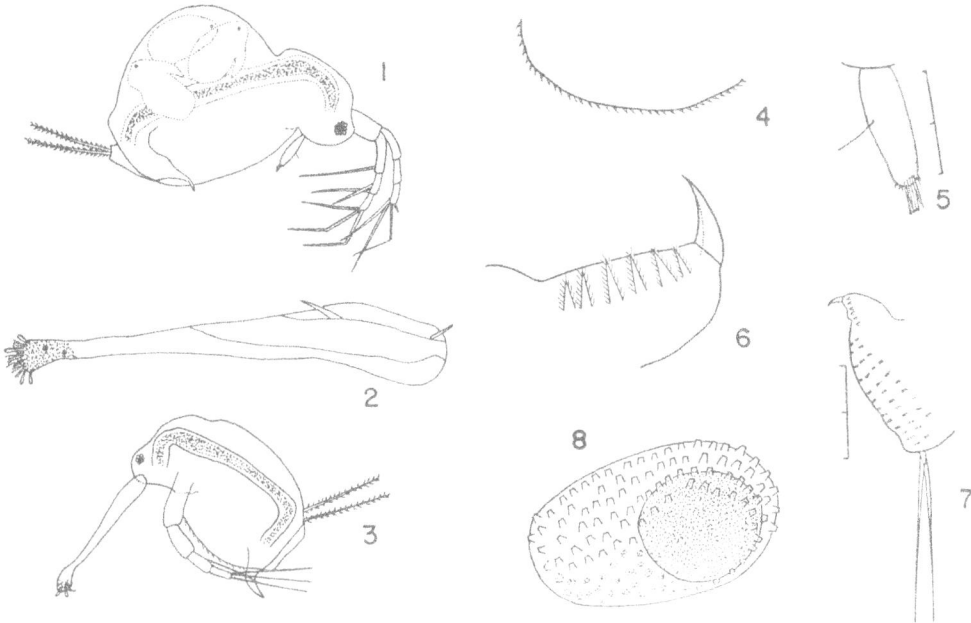

FIG. 41. Olivier's illustrations of *Moina eugeniae* (from Olivier, 1954; by permission of the author). 1. Parthenogenetic female.
2. Male antennule. 3. Male. 4. Female shell margin. 5. Female antennule. 6. Female postabdomen. 7. Female postabdomen.
8. Ephippium.

la cabeza. Formula antenal 0-0-1-3/1-1-3 y sin otras
características especiales. Postabdomen de tamaño mod-
erado, región post-anal corta y de terminación distal
cónica. Uña terminal ⅓ del large de la región post-anal
del postabdomen, no pectinada o tenuamente pectinada
[fig. 41,6]. Región post-anal del post-abdomen con 5–7
espinas plumosas, sin el diente bífido característico de casi
todas las especies de este género. Una serie irregular de
grupos de finos pelos se observa en la region pre-anal
[fig. 41,7]. Ephippium pustulado y con un solo huevo de
resistencia [fig. 41,8].

El largo de la hembra adulta varía en 25 ejemplares
medidos entre 0,61 mm como mímino y 1,05 mm como
máximo, mientras que la mayoría oscila entre 0,83 y
0,97 mm. El ancho que es muy variable según lleve o no
embriones la hembra, varío en los 25 ejemplares entre
0,27 mm como mínimo en ejemplares sin embriones y
0,84 mm como máximo en individuos con varios embrinoes.

Macho: El macho no es mucho más pequeño que la
hembra; su largo oscila alrededor de los 0,83 mm y el ancho
en los 0,55 mm. La cabeza posse el borde ventral menos
convexo que la hembra, formando el dorsal una curva más
suave que en la hembra, con sinus supra-ocular presente.
Borde dorsal de las valvas curvo, lo mismo que el ventral
que lleva espinitas [fig. 41,3]. Las anténulas son muy
largas, 0,50 mm, es decir más de la mitad del total del
cuerpo, y muy robustas; se insertan muy cerca del borde
ventral. For transparencia se observan poderosos mús-
culos que salen aproximanente de la mitad la anténula.
Son además suavemente arqueadas y en el tercio anterior
del borde convexo posse una fuerte espina y otra más chica
muy próxima a su base. El extremo es más bien globoso

y lleva 4–6 tentaculitos cortos y gruesos y un manojo de
cerdas sensitivas más pequeñas; es además granuloso y
posee dos manchas obscuras características [fig. 41,2].
El postabdomen proporcionalmente es menor que en las
hembras. (Olivier, 1954: pp. 81–85).

DIFFERENTIAL DIAGNOSIS

Moina eugeniae and *hutchinsoni* may easily be
distinguished from all other species of the genus *Moina*
by the absence of the bident tooth on the postabdomen. The typical form of *hutchinsoni* from Nevada
does have a unident tooth which appears to be a
remnant of the bident tooth. The Soap Lake form of
this species lacks any trace of a tooth and therefore is
similar to *eugeniae*. However, *eugeniae* is a much
smaller species than *hutchinsoni*—the latter species
measures about 1.6 mm. long while *eugeniae* may be as
large as 1.05 mm. (Olivier, 1954). Furthermore,
hutchinsoni lacks a supraocular depression and has a
very large head in proportion to the size of the body.

There is no doubt in my mind that *eugeniae* should
be considered a distinct species from *hutchinsoni*. The
similarity in form and habitat with *hutchinsoni* does
indicate a close relationship. It would appear that
eugeniae has developed from *hutchinsoni*-like species
since the latter retains a remnant of the bident tooth

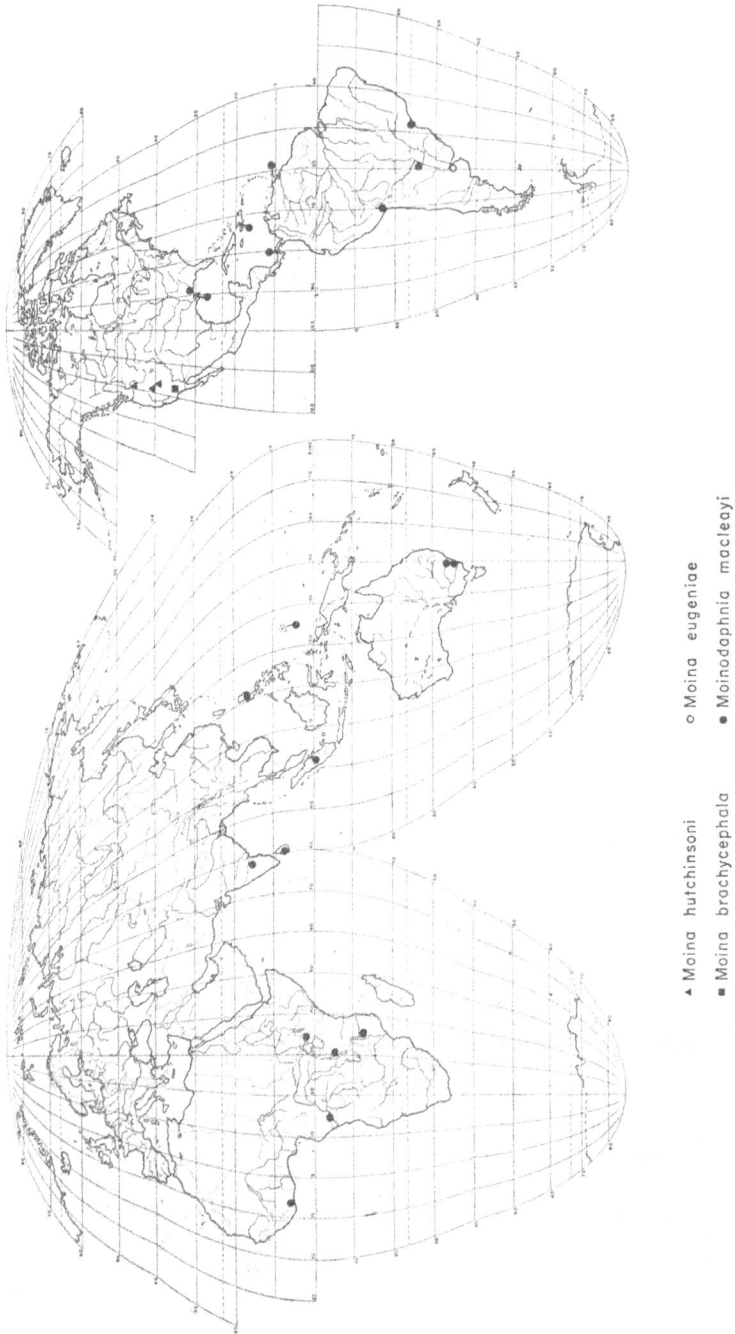

△ Moina hutchinsoni   ○ Moina eugeniae

■ Moina brachycephala   ● Moina brachycephala   ● Moinodaphnia macleayi

FIG. 42. Geographical distribution of *Moina hutchinsoni*, *eugeniae*, *brachycephala*, and *Moinodaphnia macleayi*.

and the latter has a two-egged ephippium which appears to be a "primitive" feature in *Moina*.

DISTRIBUTION

The type locality for this species is the Laguna Salada Grande in Argentina (fig. 42). It is also the only locality from which the species has been collected. According to Olivier the species is an euplanktonic form.

No specimens have been examined.

MOINA BRACHYCEPHALA SP.N.

DIAGNOSIS

This species is characterized by the very large and broad head that is much larger in proportion to the whole body than in any other species of *Moina*. The head is very flat along the anterior and dorsal margins. There is no supraocular depression or hairs on the head. The eye is quite small and lies in the middle of the head. The antennules of the female are long and

FIG. 43. *Moina brachycephala* sp.n. *A.* Female postabdomen. *B.* Sexual female (ephippium not illustrated). *C.* Antennule. *D.* Female first leg. All specimens from type locality.

thin. The second antennae are pubescent but completely lack the vertical row of teeth on the four-segmented ramus.

The postabdomen lacks long hairs on the dorsal margin. There are ten to twelve lateral feathered teeth and one short bident tooth on the postabdomen itself, and a distinct pecten on the claw.

The ephippium is ornamented with a cobblestone pattern of protuberant round cells.

The male has a large head and a small eye. The antennules are very long and are bent near the head. In addition to the three terminal hooks on the distal end of the antennule, there are three accessory hooks arranged in a vertical row on the medial margin.

The first leg of the male has a rather small hook and a moderately long exopod segment. The vas deferens opens laterally near the base of the postabdominal claw. The spermatozoa are small spherical cells.

DESCRIPTION

The head of *Moina brachycephala* is very large, at least one-third of the total body length (fig. 43, *B*). The specific name derives from the very broad head which is flat in front and on the dorsal margin. There is a ridge that forms a broad circle dorsal to the second antenna which tends to delimit the lateral margins of the head. There is no indication of a supraocular depression although the cephalic muscles are attached to the inner margin of the exoskeleton as in the other species.

The eye is extraordinarily small and lies near the middle of the head just below the intestine and the pair of gastric caeca.

The antennules are connected below the eye and are long and thin, almost as long as those of *tenuicornis*. The antennules have a long sensory seta that originates from a point approximately one-third the length of the antennule from the head (fig. 43, *C*). There are many setae scattered around the antennule, but these are not arranged in rings. There is also a vertical row of long hairs on the lateral margin.

No hairs have been observed on either the head or shell.

The second antennae are of a usual shape for the genus but are not so pubescent as in most of the species. The vertical row of teeth on the four-segmented exopod is reduced to a few teeth near the distal end of each segment.

The shell is not as large, in proportion to the head, as in other species of the genus. The shell surface is reticulated. The ventral shell rim has thirty to thirty-five long setae that are followed behind by a row of very short, equally sized setae. The shell hooks at the dorsal corner of the posterior margin of the shell are rounded, not elbow-shaped.

The postabdomen is similar to the postabdomen of other species of *Moina* and is ornamented along the dorsal margin with rows of short setae, but there are no long hairs. The lateral teeth, from ten to twelve in number, are feathered with heavy spines, not with hairs as is usually the case (fig. 43, *A*).

The claw has a "Basaldorn" of four to five thin teeth and a pecten of thirteen to fifteen rather strong and long teeth. These teeth are almost equal in length to the width of the claw. The distal part of the claw is armed with a row of short sharply pointed teeth.

The bident tooth is short, quite similar in length to the bident of *macrocopa* and *belli*; and the proximal arm is almost as long as the distal arm.

The sexual female is frequently smaller than the parthenogenetic female as is also the case in *macrocopa*. The latter are about 1.4 mm. long while sexual females are about 1.3 mm. long.

The two-egg ephippium is ornamented with round cells that are knoblike and present a cobblestone pattern. The ephippium is brown.

Males

The males have a surprisingly large head that is much larger than the head of any other male of *Moina*. It is thus a very conspicuous feature of this species (fig. 44, *A*). The head and shell completely lack the hairs so common on the large species of *Moina*. The eye is small and lies near the middle of the head. It does not fill the tip of the head as does the eye of males in most other species of *Moina*.

The first leg has a large hook, larger than the hook of *wierzejskii*, but not larger than the first leg hook of *macrocopa* or *australiensis* (fig. 44, *C*). In addition to the usual setation on this foot, there is an exopod with a seta. This seta does not extend to the posterior margin of the shell.

The shell is not rectangular in form as is true of many males in this genus, but instead is oval and almost pointed on the posterior margin. The setae on the ventral shell rim are longer than usual. There are about thirty of these setae.

The male postabdomen is similar to that of the female except that the teeth in the pecten are smaller. The opening of the vas deferens has not been seen but appears to be near the ventral margin of the post-abdomen close to the distal end.

The antennules of the male are unique in this species (fig. 44, *B*). The bend is near the head, and the two sensory setae originate near the bend. The distal part of the antennule is longer in proportion to the body length than in most other species of *Moina*; and in addition to the three terminal hooks, there are three hooks in a vertical row positioned along the medial side of the antennules.

The type material of this species has been placed in the United States National Museum—Holotype 123200; Paratypes 123201 and 123202.

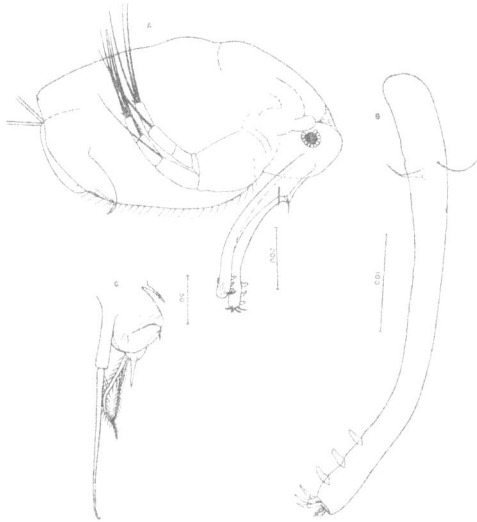

FIG. 44. *Moina brachycephala* sp. n. *A.* Male. *B.* Male antennule of *A.* *C.* First leg of male. All specimens from type locality.

DIFFERENTIAL DIAGNOSIS

This species may at once be identified by its very large head. There are two species of *Moina* that occur in the same geographical area as *brachycephala*, which might be initially confused with this species. They are *Moina macrocopa* and *wierzejskii*. *Moina macrocopa* is distinguished from all other species of *Moina* by the toothed seta on the first leg of the female. *Moina wierzejskii* usually has a very large pecten on the claws (the size varies to some extent) and has hairs on both the head and shell. These hairs are completely absent on *brachycephala*. The males of *brachycephala* are very characteristic and should not be confused with either of the other two species.

DISTRIBUTION

This species was found in a plankton collection of S. F. Light which is now stored in the United States National Museum. The collection was made by a Dr. Nickelbacher. The locality listed was "Needles to Barstow Highway, Central San Bernardino Co.; about 26 mi. S. E. of Barstow." This locality would be in the central part of the Mojave Desert in Southern Californnia (fig. 42). The collection date was April 16, 1936.

Unfortunately nothing is known about this locality. It is no doubt a very ephemeral habitat and may be slightly alkaline or saline.

IV. CHARACTERISTICS OF THE GENUS MOINODAPHNIA

The first species belonging to the genus *Moinodaphnia* was described in 1853 by King, but the genus was not described until 1887 by Herrick. As usual the generic description was brief:

Head strongly arched above, angled in front, with almost a beak behind, antennules long, movable, as in Moina, antennae with a long, unjointed spine on the apical joint of four-jointed ramus, body quadrate, merely slightly angled above; post-abdomen long, with the bifid spines characteristic for Moina, above provided with two evident processes for the occlusion of the brood sac. (Herrick, 1887: p. 35.)

Moinodaphnia may readily be distinguished from *Moina* by the presence of an ocellus, the unusual setation on the second antennae, and the well-developed fold on the abdomen that serves to close off the brood pouch. The females are laterally flattened, and the dorsal margin of the shell has a sharp keel.

The second antennae are very characteristic because the distal segment of the exopod has four rather short setae rather than three long setae. This is due to an elongation of the lateral spine of this segment and a coincidental shortening of the three swimming setae.

The ephippium of the sexual females covers a large part of the female's shell but is heavily reticulated only over a rather small area directly covering the egg locule. There are a few protuberant cells restricted to the center of the shell. Only one sexual egg is normally deposited in the ephippium.

The males have the usual long curved antennules used for clasping the female. However, although the thin sensory seta is located about one-third the length of the antennule from the head, the second stout seta is on the distal end of the antennule. There are no hooks at the distal end, rather the antennule has a row of four to six small setae on a rounded knob here.

The first leg of the male lacks an exopod and has a stout hook on the penultimate segment.

Seven species have been described as either belonging to the genus or have subsequently been placed in the genus. These are:

Moinodaphnia macleayi (King, 1853)
Moinodaphnia submucronata (Brady, 1886)
Moinodaphnia alabamensis Herrick, 1887
Moinodaphnia Mocqueryst Richard, 1892
Moinodaphnia brasiliensis Stingelin, 1904
Moinodaphnia reticulata Daday, 1905
Moinodaphnia juanae (Brehm, 1948)

Moinodaphnia brasiliensis appears to be a synonym of *Moina minuta* (see above, p. 62), while *reticulata* has been removed to the genus *Moina* (see above, p. 74). All other forms seem to be synonyms of *Moinodaphnia macleayi*. *Moinodaphnia alabamensis* is usually considered a distinct species because of its very large size. However, it has only been collected by Herrick (1887) and has never subsequently been found even though *Moinodaphnia macleayi* has frequently been reported from the same region. It seems rather likely that the length of *alabamensis*, as given by Herrick, is exaggerated.

MOINODAPHNIA MACLEAYI (KING, 1853)

Moina Macleayii King, 1853: pp. 251–252; pl. 5.
Moina submucronata Brady, 1886: p. 294; pl. 37, figs. 4–5.
Moinodaphnia alabamensis Herrick, 1887: pp. 35–36; pl. 3; figs. 13–16.
Moinodaphnia Mocquerysi Richard, 1892: pp. 222–226, figs. 7–8.
Moinodaphnia Macleayii Sars, 1901: pp. 16–19; pl. 3, figs. 1–10.
Moinodaphnia Macleayi Daday, 1910: pp. 143–144; pl. 8, figs. 7.
Moinodaphnia macleayii Birge, 1918: p. 703; fig. 1089.
Moinodaphnia macleayii Henry, 1922: p. 34: pl. 4, fig. 6.
Moinodaphnia Macleayi Brehm, 1933: pp. 669–670; fig. 10.
Moinodaphnia Macleayi Ueno, 1936: p. 515; fig. 2, A–D.
Moina Juanae Brehm, 1948: pp. 99–100; fig. 2, A–C.
Moinodaphnia macleayi Brehm, 1953a: pp. 324–325; fig. 92.

DIAGNOSIS

Head sub-triangular; eye large and fills end of head. An ocellus is present. The second antennae have long rami and are with a long sensory seta between the two rami. The distal end of the terminal segment of the exopod has a spine, though normally short in all species of *Moina*, is lengthened and of equal length to the three distal feathered swimming setae in *Moinodaphnia*. The shell is round and has a slight crest on the dorsal margin. The ventral shell rim has a row of small setae. Postabdomen of typical form. The claw is not pectinate. A horseshoe-shaped fold is present on the dorsal side of the intestine that serves to hold embryos in the brood pouch.

The male's long antennules are bent at a point one-third the distance from the head. The distal end of the antennule has six long hooks. The first leg is with a well-developed hook but no exopod. The post-abdomen has a large claw pecten.

DESCRIPTION

This species was first described by King (1853) from Australia. Several subsequent forms have been described, and all of these forms, including the present species, have been placed in a distinct genus, *Moinodaphnia*. It seems apparent that these forms all represent the same species, *macleayi*, and that this

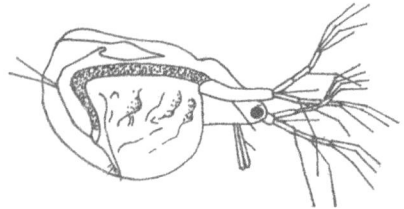

FIG. 45. King's original illustrations of *Moina macleayii* (= *Moinodaphnia macleayi*; from King, 1853).

species does deserve generic separation from the genus *Moina* on a morphological as well as an ecological basis.

King (1853) described *macleayi* as follows (fig. 45):

Head triangular; the eye in the apex of the triangle, large; carapace roundish-oval, smooth, without setae on the margin. The superior antennae are long, with a single seta springing from the upper edge. The inferior antennae have the basilar joint of moderate size; a single seta springs from one of the crenations on the side, and a jointed seta, nearly as long as the posterior branch of the antennae, springs from the top. The anterior branch has four short plumose setae springing from the extremity of the last articulation, and one from the penultimate. The posterior branch has three short-jointed setae and a short spine springing from the extremity of the last articulation —one longer from the penultimate, and one still longer from the antepenultimate. The last segment of the abdomen is longer than in *M. lemnae*, and has one small process directed upwards, closing the receptacle for the ova. There are small spines round the anus. (King, 1853: p. 251.)

Female

The shell is broadly rounded while the small head is triangular in shape. The eye is large and fills the tip of the head. There is a slight supraocular depression above the eye, but this is not too obvious. An ocellus is present and is located above the origin point of the antennules. The ventral margin of the head is flat, not rounded; and the antennules arise along this margin and just behind the eye. The antennules are long and thin, lack the vertical row of long hairs, and are ornamented only with horizontal rows of short setae. The sensory seta is very long and originates along the anterior margin at the mid-point. The distal end of the antennule has nine sensory papillae, not four feathered papillae as described by Herrick (1887: p. 35).

The labial keel is of the usual form and does have a few hairs on its ventral margin. There are no hairs anywhere else on the head or shell.

The second antennae are thin and long. The basipod segment has two sensory setae that stem from the posterior margin near the base and one long sensory seta that originates at the distal end of the basipod and between the two rami. This latter seta is almost as long as the rami. The three-segmented endopod has the normal pattern of swimming setae (fig. 46, *A*), but the exopod has the distal segment with four setae. One of these four setae represents the normally reduced seta found on all species of *Moina*. In *Moinodaphnia* this seta is very long and is equal in length to the three feathered setae that, however, are somewhat shorter than usual. The exopod also lacks the vertical row of teeth on its inner margin.

The valves of the shell come together at the mid-line of the body axis in adult specimens, and the posterior half of the animal is expanded in a dorsal direction and has a slight keel. This form of the shell is similar to that found in adult specimens of *Simocephalus* and several of the macrothricids. Immature specimens, however, may have the valves joined at the dorsal end of the posterior margin, and the dorsal margin may be rather straight—not so curved. The above feature which *Moinodaphnia* shares with *Simocephalus* and some macrothricids seems to be an adaptation for a benthic or littoral life. According to Sars (1901) *Moinodaphnia* may often be seen stationary, with its back to vegetation. This is a typical position for *Simocephalus*, and in these two forms, it is probably a parallel development (*Moinodaphnia* would seem not to have any association with the daphniids because its male is strikingly different and lacks the exopod on the first leg).

The shell is reticulated. The ventral shell rim has a row of short setae along its full length. In front, these setae appear to be distinctly separated from the shell and therefore would appear to be comparable to the long setae on the shells of *Moina*. In *Moinodaphnia* the setae are highly reduced in length. There are also groups of fine hairs between the shell setae. There is a pair of submarginal hooks at the point where the valves come together.

The first leg of the female is like that of *Moina* females. The two anterior setae are short, but the middle distal feathered seta is very long. The fifth leg appears to be well developed and has an elongated base.

The brood pouch is almost completely closed off by a horseshoe-shaped fold that rings the dorsal and lateral sides of the abdomen. This fold is similar to the abdominal process found in the daphniids although here it is a broad fold while in the daphniids the process is a fingerlike projection. The "Nährboden" seems to be well developed in this genus.

The postabdomen is much like that of *Moina* although the distal part is more tapered. There are ten to eleven feathered teeth and one bident tooth on the lateral sides. The bident tooth has the two arms rather widespread and almost of equal length.

The claw lacks a pecten, but there is a row of fine hairs along its full length.

The dorsal margin of the postabdomen is ornamented with the fine setae pattern as found in *Moina*. There are no long hairs present.

Females measure .9 to 1.1 mm. long.

The ephippium of the sexual females covers only a small part of the shell, is heavily reticulated in the center (fig. 46, *A* and *B*), and laterally expanded in contrast to the shell of the parthenogenetic female (fig. 46, *C*).

Male

The head is elongated as in males of *Moina* and has a large eye that fills the end of the head (fig. 46, *D*). There is an ocellus present. The antennules are long and curved inward. The antennules are as long as those of the males of *Moina* in relation to body size. The thin sensory seta is on the lateral margin at a point approximately one-third the length of the antennule from the head. The second seta originates near the distal end of the antennule and on the medial margin (fig. 46, *F*). Immediately proximal to this seta is a round knob that carries a row of four to six short setae. The sensory papillae are at the distal tip.

The setation of the second antenna is like that of the females and has four swimming setae rather than three on the distal end of the exopod.

The shell is broadly rounded along the ventral margin and has an indistinct posterior margin. The ventral margin has a row of short setae.

The first leg lacks the exopod. There is a large hook on the penultimate segment, and the ultimate segment is reduced and lies medial to the hook (fig. 46, *E*). This latter segment bears three setae; two are feathered while the third is curved like a hook. The first leg is almost identical to that of *Moina micrura*.

The postabdomen is like the female's postabdomen. The precise location of the sperm duct opening has not been observed, but it seems to be on the ventral margin near the fold between the postabdomen and the abdomen. The spermatozoa are small round cells.

The males measure .55 to .60 mm. long.

Although the female of *Moinodaphnia* bears little resemblance to any of the species of *Moina*, the male shares some characteristics, particularly the form of the first leg, with *Moina micrura* and *brachiata*. However, it is probable that *Moinodaphnia* has been separated from the main moinid line for a considerable time.

It does not seem possible to consider *Moinodaphnia* a "missing link" between *Moina* and *Daphnia*, as Herrick (1887) did because it has lost many features shared by the more generalized species of *Moina*, the sidids, and the daphniids. *Moinodaphnia* lacks an exopod on the first leg of both the female and the male. This exopod is present in all of the daphniids, sidids, and in males of the larger species of *Moina* such as *macrocopa* and *tenuicornis*. Furthermore, the ephippium of *Moinodaphnia* contains only one sexual egg; whereas two eggs seem to be the primitive condition in *Moina* and in the daphniids. *Moinodaphnia* instead must be considered a highly specialized form. The presence of an ocellus and of the large horseshoe fold on the postabdomen to protect the embryos would seem to have been secondarily acquired.

The type material of *M. submucronata* (Brady, 1886) and *M. mocquerysi* may be found in the British Museum (Natural History).

DIFFERENTIAL DIAGNOSIS

Moinodaphnia macleayi can easily be distinguished from *Moina* by its laterally flattened body, the presence of an ocellus, and by the four setae, rather than three, on the exopod of the second antenna. The two species that have a rather similar body morphology are *Moina minuta* and *reticulata*; but these species

have a very different postabdomen and only three swimming setae on the second antenna. Furthermore, *minuta* has an unusual first leg that lacks the anterior seta on its penultimate segment.

DISTRIBUTION

Moinodaphnia apparently has a completely different distribution and habitat from that of *Moina*. Rather than restricted to small temporary ponds, *Moinodaphnia* is found in small lakes, swamps, and pools. It is not a zooplankter but instead lives near the bottom mud or in the weeds where it may attach itself by the back, like *Simocephalus*, and filter the water surrounding the weeds.

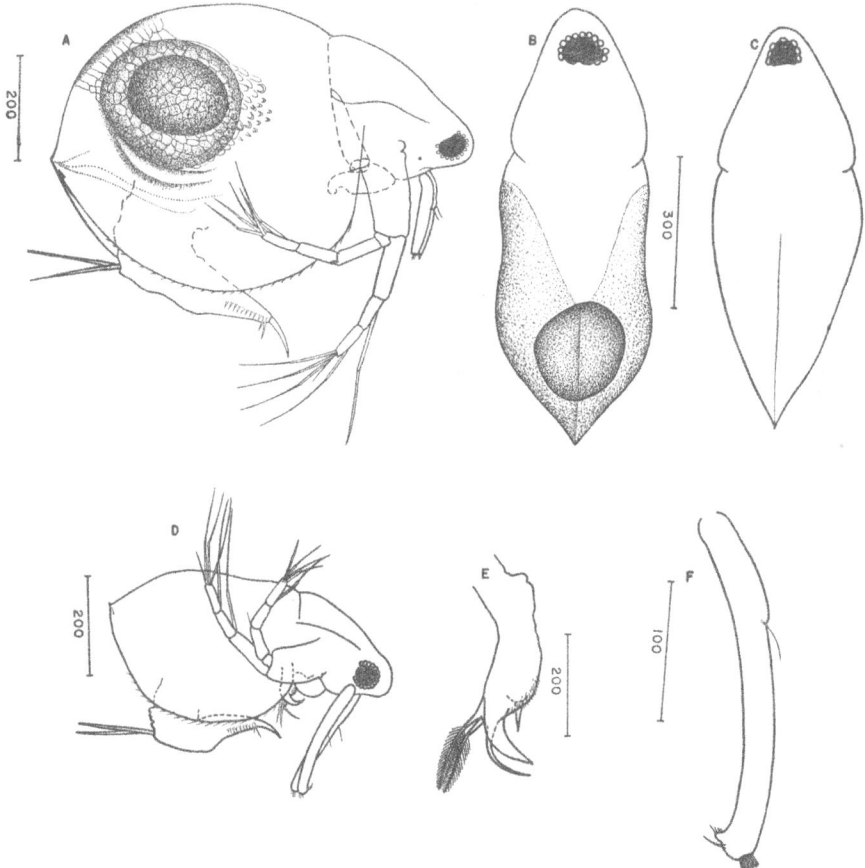

FIG. 46. *Moinodaphnia macleayi* (King). *A.* Sexual female. *B.* Dorsal view of sexual female. *C.* Dorsal view of parthenogenetic female. *D.* Male. *E.* Male first leg. *F.* Male antennule. (All specimens from ditch along Fort Portal Road, Kampala, 21-XI-56, Uganda; collected by I. F. Thomas.)

Moinodaphnia is very widely distributed throughout the humid tropics (fig. 42). It has been reported throughout the central part of Africa, Southeast Asia, India, Ceylon, Australia, on several Pacific Islands, South America, the Caribbean Islands and the southern United States.

Specimens Examined

North America:
Louisiana:
New Orleans, 1904 (Birge Collections, slide No. 4–6)
Central America and West Indies:
Haiti:
Tron Caiman, 16-II-33, collected by R. M. Bond
Panama:
No locality given (Birge Collection, slide No. 952)
Trinidad:
No locality given (Birge Collection, slide No. 473)
Africa:
Congo Republic:
Caca Mueca near Brazzaville, 10-VI-1890, (Birge Collection slide No. 146—determined by Richard as *M. mocquerysi*)
Uganda:
Fort Portal Road, Kampala, 21-XI-56, collected by I. F. Thomas
Southeast Asia:
Ceylon:
Colombo, collected by A. Haly (determined by Brady as *Moina submucronata*), British Museum, Natural History, Collection No. 1951.8.10.757)
Philippines:
No locality given (Birge Collection, sample No. 5)

V. EPHIPPIUM FOSSILIS

A fossil ephippium of *Moina* was found by Allison Palmer in a calcareous nodule from the Barstow Formation, Southern Calico Mountains, of the Mojave Desert in California (USGS Locality 19057, California, CMC-7). The deposits are presumed to be of a Late Miocene age (Palmer, 1957). The nodule was from a locality where an anostracan, two arachnids, and several insects have been found (Palmer, 1957) and is near the locality where fossil copepods were found (Palmer, 1960). Although little is known of the habitat, it appears that the area must have contained a rather large lake.

The ephippium (fig. 47) is ornamented with polygonal reticulations around the edges and is indistinctly reticulated in the middle. From the structure it obviously would contain one, rather than two, sexual eggs. The ephippium is 525 microns long and 343 microns high.

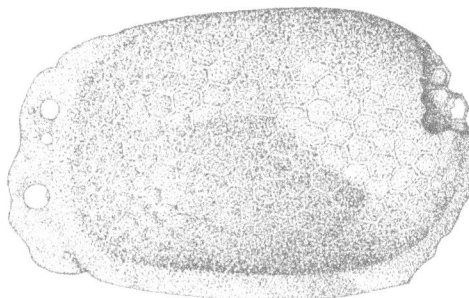

Fig. 47. Fossil ephippium of Late Miocene age from Barstow Formation, Southern Calico Mountains, Mojave Desert, California (illustrated by W. Vars).

In all aspects this ephippium is similar to that of *Moina micrura*, and I believe that it is either a closely allied form or represents the same species.

VI. FUNCTIONAL MORPHOLOGY OF THE SPECIES CHARACTERS

The morphological characters which differentiate the species of the Moinidae fall broadly into two categories: (A) characters of ecological significance, and (B) secondary sex characters.

A. MORPHOLOGICAL CHARACTERS OF ECOLOGICAL SIGNIFICANCE

Although all of the species exhibit the basic morphology which was described earlier (see above, pp. 6–10), individual species may be distinguished by body size, head form, and the presence, density, and pattern of setae on the body and appendages. There is also an interspecific variation in the number of sexual eggs deposited in the ephippium, although this variation seems ultimately to be related to body size.

BODY FORM

Body size

Table 1 gives the size range of the mature females of each species. There is obviously considerable variation, the smallest species—*micrura*, *minuta* and *flexuosa*—measuring as little as 0.5 mm., while the largest—*mongolica* and *wierzejskii*—may reach lengths of 1.7–1.8 mm. Why the species of this group should differ so much in size is difficult to ascertain. It will be seen in the phylogeny, however, that the smaller the species are the more advanced, and that, therefore, the trend in moinid evolution has been towards reduced size. Of the factors which would be likely to increase selective pressure in this direction, predation seems to be the most obvious. Insects and fish feed heavily on the Cladocera, but predation is most severe on populations of large, easily visible, species (Hrbaček, 1961;

TABLE 1

Variability of Morphological Characters of Parthenogenetic Females of the Moinid Cladocera

	Length (mm.)	No. sexual eggs	Supra-ocular depression	Ocellus	Hairs on head	Hairs on shell	No. setae on ventral shell rim	Pattern of setae on posterior shell rim	Shell hooks	Teeth-like setae on 2nd ant. exopod	Abdominal fold	Anterior seta on ♀ 1st leg	Hairs on postabdomen	No. feathered teeth on postabdomen	Distal bident tooth
M. brachiata	1.2–1.6	1	P	A	Patch behind 1st Ant.	A	35–40	grouped	P	P	A	P	A	9–14	well dev.
M. macrocopa	1.0–1.5	2	A	A	P	P	55–65	ungrouped	P	P	A	Toothed	P	7–10	well dev.
M. micrura	.5–1.2	1	P	A	A	A	11–25	grouped	P	P	A	P	A	3–11	well dev.
M. affinis	.8–1.2	1	P	A	P	P	20–27	ungrouped	P	P	A	P	P	7–14	well dev.
M. wierzejskii	1.0–1.7	2	A	A	P	P	20–26	grouped	P	P	A	P	P	9–12	well dev.
M. australiensis	1.2–1.5	2	P	A	P	P	ca. 30	ungrouped	P	P	A	P	P	10–12	well dev.
M. tenuicornis	ca. 1.2	2	A	A	P	P	ca. 40	grouped	P	P	A	P	P	10–12	well dev.
M. weismanni	.9–1.0	1	P	A	P	A	ca. 17	grouped	P	P	P	P	A	7–9	both spines
M. flexuosa	.7–.9	1	P	A	A	A	ca. 18	ungrouped	P	P	A	P	P	6–7	well dev.
M. hartwigi	.8–1.2	1	P	A	Patch behind 1st Ant.	A	23–27	ungrouped	P	P	A	P	P	7–8	well dev.
M. minuta	.5–.7	?	P or A	?	A	A	13–16	grouped	P	A	A	A	A	3–6	well dev.
M. mongolica	1.0–1.8	1		A	A	A	29–30	usually ungrouped	A	A	A	A	A	7–10	small
M. belli	1.2–1.7	2	A	A	P	P	40–60	ungrouped	P	P	A	P	P	5–7	well dev.
M. reticulata	.6–.7	?	P	P	A	A	13–16	ungrouped	A	A	P	P	A	5–6	both spines
M. hutchinsoni	1.4–1.6	2	A	A	P	P	ca. 40	ungrouped	A	P	A	P	A	7–9	well dev.
M. eugeniae	ca. 1.0	1	P	A	?	?	?	?	?	?	?	?	?	5–7	A
M. brachycephala	1.3–1.4	2	A	A	A	A	30–35	ungrouped	P	A	A	P	A	10–12	both spines
Moinodaphnia macleayi	.9–1.1	1	P	P	A	A	35–40 that extend onto posterior rim		P	A	P	P	A	10–11	well dev.

(P = present; A = absent)

Brooks, 1965; Brooks and Dodson, 1965, discuss the relative affects of predation on large and small species of Cladocera). Although moinids are commonly found in small temporary habitats where large predators are absent, it must be remembered that the smaller species also live in permanent environments such as ponds and lakes.

Head Morphology

In the smaller species of *Moina* the head is narrow, depressed, and with a distinct supraocular depression, while the larger species have a broadly rounded head which lacks the supraocular depression.

The supraocular depression is formed by the attachment of a muscle bundle from the intestine and labrum. Constriction or shortening of the muscles would produce an indentation and also (and probably more important) depress the head. The latter may have developed in response to a different swimming posture, for the smaller species swim in a more upright position than the larger species. This would be advantageous to the animal because a depressed head would permit a better view of objects below the animal.

SETAE PATTERNS

Most of the species of *Moina* and *Moinodaphnia* have combs of setae on the second antennae, first leg, postabdomen, and shell, which may be used in cleaning the body of silt and detritus. Since these setae are grouped or are restricted to only part of the body and since they vary in number and size, they may be used as species characters.

Setae on the Shell

The surface of the shell and the head are often covered with thin hairs. In the most specialized forms of the group (*Moina micrura, reticulata,* and others) these hairs are absent, and their distribution in other species is highly variable. In *Moina brachiata* and *hartwigi* the hairs are restricted to a small area behind the antennules; in *weismanni* the hairs occur on the dorsal surface of the head and the anterior half of the shell, while in several other species, including *macrocopa, wierzejskii,* and *tenuicornis,* the hairs are found over the entire surface of the shell. That they are used to keep silt and clay off the body is borne out by their being most common on species that regularly inhabit turbid, temporary ponds.

The shell is also ornamented with rows of setae along its anterior and ventral margins and with a row of fine hairs on the posterior margin. The number of setae on the ventral rim varies considerably between species (as does also the pattern of fine hairs on the posterior rim). In *Moina micrura* and *minuta* there may be only fifteen to twenty setae, while in *macrocopa* there are fifty to sixty.

The fine hairs on the posterior rim may be arranged into two different patterns. The hairs may be of equal size and very numerous (*affinis, wierzejskii, macrocopa, tenuicornis, flexuosa, brachycephala,* and *belli*), or they may be organized into groups or clusters of six to eight (*micrura, brachiata, australiensis, minuta,* and *weismanni*). Each cluster is composed of a graded series of hairs that increase in size posteriorly (fig. 25, *B*).

The setae and hairs on the shell rim may help to clean the postabdomen as it moves back and forth between the shell valves. There is a clear relationship between the loss of hairs on the postabdomen and the increased grouping of the fine hairs on the posterior shell rim.

Setae on the Second Antennae

The second antenna is composed of a basal segment, the basipod, and two distal rami, the exopod and endopod both of which bear the swimming setae. The segments of each component of the antenna are covered with rings of short setae and with vertical rows of long hairs that are restricted to the medial margin of the ramal segments. The exopod also bears a vertical row of stout teethlike setae along the inner margin (fig. 15, *B*) of each segment. These extend the full length of each exopod segment only in species found commonly in turbid waters, but are completely absent in those species, including *Moinodaphnia macleayi, Moina mongolica,* and *Moina reticulata,* which live in relatively clear permanent ponds or lakes. Their function is probably that of cleaning silt from the shell surface when the second antenna is scraped against the female's body.

Two other features of the second antennae that differ among species are the comparative length of the sensory setae at the base of the basipod, and the length of the spine on the distal segment of the exopod. In those species that are planktonic in large freshwater lakes, *minuta* and *micrura,* the two sensory setae on the basipod are very long. Hantschmann (1961) suggested that these setae detect the flow of water past the body and thereby may indicate the sinking rate of the suspended animal.

In all species of *Moina* the spine on the distal exopod segment is very short. However, in *Moinodaphnia* it is quite long, almost as long as the swimming setae. *Simocephalus* has a comparable spine on the second antenna (in this case a modified swimming seta) that serves to support the body when the animal rests with its back to the substrate. *Moinodaphnia* has a similar habit and no doubt the modified spines act in the same manner.

First leg

The distal and penultimate segments of the female first leg each have a seta on the anterior margin.

Normally these two setae are bare, without the fine hairs as on the other setae of this leg (fig. 43, *D*). However, in *Moina macrocopa* these two anterior setae are toothed on their outer margin. The teeth are so situated that they scrape the inner shell surface and part of the second filtering limb and thus may clean silt and other debris.

In two species of *Moina*—*mongolica* and *minuta*—the anterior seta on the penultimate segment is absent. These two species are commonly found in large permanent lakes that are less turbid than temporary pools.

Postabdomen

The postabdomen of *Moina* is divided into a proximal broad base and a distal conical postanal extension (fig. 43, *A*). The dorsal margin of the proximal base is ornamented with rows of short setae, and in some of the species long hairs are also present. The presence or absence of these long hairs may be used as a species character. Although their precise function is unknown, they probably help clean silt from the posterior part of the inner shell surface or, like the fine hairs on the shell, prevent the accumulation of silt on the postabdomen.

The distal conical extension of the postabdomen is armed with a distal claw and with a lateral row of teeth on either side. The claw apparently aids in removing excess food or detritus from the anterior end of the food groove (Fryer, 1963). The lateral teeth, unlike those of most Cladocera, are feathered with fine hairs (fig. 11, *A–F*). The distal-most tooth of this series is not feathered and is a bident tooth. The lateral teeth seem to be particularly well suited for cleaning detritus from the filtering combs of the trunk limbs, and the distal bident tooth may serve the same purpose. This bident tooth is absent in *Moina hutchinsoni* and *eugeniae* and is considerably reduced in size in *mongolica*, the three species specifically adapted to live in very saline lakes.

The pair of long setae that stem from the dorsal margin of the postabdomen project out between the posterior margins of the shell and probably aid in keeping embryos in the brood pouch. In *Moina* these setae are commonly caught at the dorsal end of the posterior shell margin by a pair of hooks projecting from the shell. When caught in these shell hooks, the setae remain in a stationary position and block the posterior end of the brood pouch. In species with an open brood pouch, the posterior opening would of course need to be closed off in order to avoid loss of the developing embryos. The shell hooks are present in all species except *Moina reticulata* and the three saline species *mongolica, hutchinsoni,* and *eugeniae.* In *reticulata* there is a well-developed abdominal fold that serves to retain the embryos while in the three saline species a closed brood pouch has probably developed in response to the external saline environment. In

hutchinsoni this has been formed by the shell pressing against the sides of the abdomen. When gravid *hutchinsoni* females are pressed with a pin, embryos do not escape although the same procedure in other species dislodges embryos from the pouch. A comparable closed pouch has probably developed in the other two saline species of which, unfortunately large collections were unavailable for study.

B. SECONDARY SEX CHARACTERS

EPHIPPIUM

The ephippium is produced by the shell of the female and consists of thickened pigmented cells that bulge above the shell surface. As distinguished by the shape of the individual cells, there are two types of ephippia: one composed of rectangular or polygonal cells, the other of round cells. These two major types can then be subdivided into five subgroups.

In those species possessing ephippia characterized by rectangular or polygonal cells, three subgroups may be distinguished on the basis of reticulations. The most generalized type, characteristic of *Moina tenuicornis, macrocopa,* and *belli,* is a two-egg ephippium with a completely reticulated surface (fig. 6, *A*). *Moina brachiata, micrura,* and *mongolica,* which have one-egg ephippia, show reticulations that are distinct around the periphery but indistinct (e.g., *micrura* and *mongolica*) or absent (*brachiata*) on the embossed globe which forms the ephippial center (fig. 9, *E, F*). The third subgroup (*Moinodaphnia macleayi*) shows polygonal reticulations on only part of the shell, with a group of knoblike cells in the center (fig. 46, *A*).

Among species whose ephippia are composed of round cells, we can distinguish those where the cells are relatively small and project well above the shell surface from those with larger cells which tend not to protrude as much. The former is represented by *australiensis, brachycephala, wierzejskii,* and *hutchinsoni* among species with two-sexual eggs, and *eugeniae* and *flexuosa* with only one-sexual egg. Only two species, *affinis* and *weismanni,* form the latter subgroup (fig. 15, *A*).

Males

Antennules of the male. The male antennule consists of two parts, a proximal base, and a distal curved portion that is modified for clasping the female. Two sensory setae, a lateral, thin seta and a medial thick seta, originate at the knee or point that separates these two parts of the antennule.

In *Moina wierzejskii, hutchinsoni,* and *affinis,* the sensory setae are located at the base of the antennule near the head, while most of the antennule is broadly curved. Several other species, including *australiensis* and *tenuicornis,* have a rather similar antennule, but with the sensory setae originating at a point about one-fifth the length of the antennule from the head.

In other species, particularly in *brachiata*, *hartwigi*, *micrura*, and *mongolica*, the sensory setae are at a more distal point, about one-third the length of the antennule from the head. In *Moina belli*, a slight variation of this pattern is found. The medial seta originates as in the latter group, and there is a sharp bend in the antennule here, but the lateral seta originates at the mid-point of the antennule. In *macrocopa* both sensory setae are found at the mid-point of the antennule, and the antennule is bent at this point.

Finally, in *Moinodaphnia* the lateral seta originates about one-third the distance from the head while the medial seta is at the distal end of the antennule.

First Leg of the Male

The first legs of the males are modified to aid in holding the female during copulation. The penultimate segment has a hook produced by an outgrowth of the segment—it does not appear to be a modified seta. The distal segment with its three setae is considerably reduced in size and one of these setae is curved like a hook.

In most of the large species the first leg of the male has retained the exopod that has otherwise been lost in the females of *Moina* and *Moinodaphnia*. This exopod consists of only a single segment with a long distal seta. This seta has a sharp bend at its tip which probably hooks onto the female during copulation. Only *Moina wierzejskii* and *hutchinsoni* of the two-sexual egg species lack this exopod, while all species with one-sexual egg lack it.

The size of the hook on the penultimate segment varies—in some species it is considerably reduced and is then nonfunctional. *Moina tenuicornis* and *australiensis* have a generalized first leg where the hook is rather large, and the exopod has a long distal seta. In *Moina belli* and *macrocopa* the hook is larger, while the distal seta of the exopod is quite long (figs. 7, *B*; 34, *D*).

All other species, except *brachycephala*, lack the exopod. Some of these forms, including *brachiata*, *hartwigi*, *mongolica*, *micrura*, and *Moinodaphnia macleayi*, have a very large hook on the endopod.

Then we have a group of species, including *wierzejskii*, *hutchinsoni*, *affinis*, *weismanni*, *flexuosa*, and *eugeniae*, in which the exopod has been lost, the leg is reduced in size, and the endopod hook is reduced to a thin fingerlike projection (fig. 19, *C*). In these same forms, however, one of the setae of the distal segment has increased in size and projects out like a finger.

The tendency to have a smaller hook on the first leg and the concordant loss of the exopod is seen in the intermediate condition in *brachycephala*, where the endopod hook and the exopod are reduced in size (fig. 44, *C*). This may be contrasted to the condition in *australiensis* and *tenuicornis*.

The differences in the size of the hook on the male's first leg and the size of the leg itself are correlated with the differences in position of the sensory setae found on the male's antennules. Among species with a large hook and a large first leg, the sensory setae of the antennule are located at a point one-third or more the length of the antennule from the head. In those species with a small hook and leg, the sensory setae are quite near the head. This relationship is not however found in *australiensis* and *tenuicornis*.

Form of the Spermatozoa and the Opening of the Sperm Duct

In *australiensis* and *tenuicornis*, the most generalized species of *Moina*, the spermatozoa are small round cells. This is also true of all species of the group typified by *brachycephala* and *wierzejskii* which have reduced hooks on the first leg of the male and have an ephippium ornamented with round cells. On the other hand, the group of species that have a large hook on the male's first leg, generally has more variously shaped spermatozoa as in *Moina belli* and *macrocopa* where the spermatozoa are rod- or sickle-shaped cells (fig. 34, *B*), or *brachiata*, *micrura*, and *hartwigi*, whose spermatozoa are round with many radiating axons (fig. 12, *C*). However, in *mongolica* and *Moinodaphnia* the spermatozoa are small round cells.

The location of the sperm duct opening has been somewhat difficult to ascertain, but in most species it appears to be located proximal and ventral to the row of lateral teeth on the postabdomen. In *macrocopa* and *belli* it is at the distal end of the postabdomen just ventral to the base of the claw.

Functional significance of the secondary sex characters

It is obvious that the features under consideration here do not have any ecological significance, yet the great variation between species indicates that they have functional significance.

The differences in development of the first leg of the male are readily seen to be correlated with the manner in which the male secures the female for copulation. Copulation has been observed and described in *Moina brachiata* and *macrocopa* by Weismann (1879) while I have observed it in *Moina affinis* and *wierzejskii*. The males of the first two species have well-developed first legs while the latter two species have poorly developed legs.

In my own observations with *affinis* and *wierzejskii*, the male holds the female with his antennules around the groove between the head and shell and just behind her second antennae. He first takes a position on the female's dorsal side but then moves to the ventral side so that the ventral shell openings of the two are apposed. The male continues to hold the female with his antennules while he spreads the female's shell apart with the fingerlike seta on the distal segment of his first legs. He then inserts his postabdomen into the female's brood pouch and ejaculates the sperm.

As described by Weismann (1879: p. 68), copulation in *brachiata* and *macrocopa* differs only slightly from that found in *affinis* and *wierzejskii*. Here, however, the well-developed hooks on the first leg are used to grasp the shell of the female. Once the male has secured the female, he usually loosens or completely releases his antennules so that the female is held only by the first legs during copulation.

This difference in behavior during copulation accounts for the different structure of the first leg. In one the legs are well developed and are used to clasp the female; in the other the legs are poorly developed and used only to spread the female's shell apart.

Variation in copulatory behavior does not, however, explain the differences in form of the male antennules or the pattern of cells on the female's ephippium. The ephippia and the male antennules differ considerably between species and in fact can be used to identify the individual species. These different patterns must be functional or else they would neither have developed nor have been maintained in each species. There seems to be only one possible explanation for the great variability observed in these structures and that is as mechanisms for the insurance of sexual isolation. Although this thesis has not been experimentally proven, I would like to suggest that the male of *Moina* is able to (a) identify a sexually mature female by contact of his antennules against her shell and (b) identify the sexual female of the proper species by the pattern of ephippial cells. In this way the species of *Moina* have developed a mechanical means of reproductive isolation.

In order to understand how such a mechanism has formed, I would like to review what is known about the sexual cycle in *Moina*. Weismann (1877b, 1879) first described the reproductive cycle in *Moina*. He found that parthenogenetic females could produce three types of eggs—those that developed into (a) other parthenogenetic females, (b) males, or (c) sexual females (the difference between parthenogenetic and sexual females is not however as rigid as Weismann believed for sexual females may reproduce parthenogenetically after production of at least one sexual brood). The parthenogenetic females could not, however, produce sexual eggs. He further ascertained that males would copulate only with adult sexual females.

The formation of the ephippial shell is completely dependent upon the presence of a developing sexual oöcyte in the ovary. It is produced simultaneously with the development of the egg and is not related in any way to fertilization (Weismann, 1877b: p. 238) being probably hormonally controlled. If the sexual egg or eggs are not fertilized, they are reabsorbed

without being deposited in the brood chamber. The completely developed ephippium is then cast off in the next molt.

The ephippium in this instance is superfluous and the unfertilized sexual female has actually wasted a considerable amount of energy to produce it. One might question the necessity of a heavily chitinized ephippium if its only purpose is to protect the sexual eggs, because the sidid Cladocera completely lack the ephippium while the bosminid and many chydorid Cladocera have only a thin shell to protect the egg. The elaborate ephippium of the moinids therefore must have some purpose other than merely to serve as an envelope for the eggs. Furthermore, why should an unfertilized sexual female expend so much energy to produce an ephippium if it is merely to be cast off?

Preparatory to copulation the male grasps the female with his antennules, thus bringing the sensory seta and the many groups of fine setae on the medial margin of his antennule into contact with the female's shell. Since the development of the ephippium is complete, contact of the large medial sensory seta against the shell should be sufficient to recognize whether or not the female is ephippial (sexual) or parthenogenetic. Furthermore, since the male antennules and the ephippia of the sexual females are species specific in form, it would seem that when the male grasps a female he can determine not only whether or not it is a sexual female but also whether it is of the proper species. It is completely possible that the male would be able to do this by the rubbing of the fine setae of his antennule against the ephippium or merely by recognizing the topography of the ephippium.

CONCLUSIONS

The morphological characters used to identify and define the parthenogenetic females of each species of moinid Cladocera are functional parts of the organism that are important in maintaining the animal as an efficient filter feeder and swimmer in many types of physical environments. With the exception of *Moinodaphnia*, these characters are not the type that would indicate niche diversification as far as food selection or behavior is concerned. They do indicate diversification in ability to tolerate turbid water although since the more generalized large species possess the dense combs of cleaning setae, tolerance to turbidity should be a primitive feature of the group. This suggests that the more advanced species have moved into new types of habitats paralleled by a decrease in body size which is apparently the result of predation (see above, pp. 87—89). There is nonetheless a considerable amount of variability and overlap of the morphological characteristics within and between parthenogenetic females of different species.

The secondary sex characters, on the other hand, are consistent and immediately characterized the species. These characters are associated with sexual behavior and appear to be important in sexually isolating the species.

A third group of characters, not dealt with here, is formed by the physiological tolerances of individual species. It would seem that these characters, like the morphology of the parthenogenetic females, vary considerably within each species. Yet in extreme examples, such as the saline species *eugeniae, hutchinsoni*, and *mongolica*, physiological adaptations would aid species definition.

Ideally, to define species one should understand the ecological and sexual behavior, physiological tolerances and growth characteristics as well as morphological differences of individual species or populations. It is difficult to do this when expediency and practical considerations require that species be easily identified from preserved specimens. Thus it is necessary to resort to minor structural differences in defining species. In the Moinidae—and this will probably prove to be true in other cladoceran groups, particularly the macrothricids—consideration of secondary sex characters (of males and sexual females) will continue to be necessary before one can be certain of identification. Identification of populations which lack sexual stages will always be difficult.

VII. PHYLOGENY AND DISTRIBUTION OF THE MOINIDAE

The Moinidae divide naturally into two distinct groups—which I shall refer to as (A) the *"tenuicornis"* and (B) the *"australiensis"* groups—each characterized by their secondary sex characters and here named from their least specialized species. The *"tenuicornis"* group includes those species whose ephippia are ornamented with polygonal or rectangular cells and whose males have a well-developed first leg while the medial sensory seta of the male's antennule usually originates at some distance from the head (only in *tenuicornis* is this seta near the head). The ephippia of the *"australiensis"* species are ornamented with round cells that are embossed above the shell surface, the male antennule has the medial sensory seta near the head, and the first leg is normally poorly developed (*australiensis* being the only exception) and the hook is nonfunctional.

Species within each group are differentiated by the cell pattern and topography of the ephippium, the number of sexual eggs per ephippium, and/or by the precise morphology of the male, but within each group species are found to share several features. Each group may therefore be divided into the following subgroups (number following each species indicates number of sexual eggs in the ephippium):

A. *"tenuicornis"* group

1. *tenuicornis* (2)

2. *belli* (2)
   macrocopa (2)

3. *mongolica* (1)
   hartwigi (1)
   micrura (1)
   brachiata (1)

4. *Moinodaphnia macleayi* (1)

B. *"australiensis"* group

1. *australiensis* (2)

2. *brachycephala* (2)

3. *wierzejskii* (2)
   hutchinsoni (2)
   eugeniae (1)
   flexuosa (1)
   weismanni (1)
   affinis (1)

Moina reticulata and *minuta* have not been placed in groups because the sexual females and males of these two species are unknown.

Since only the two-egg species retain the exopod on the first leg of the male, they must represent the ancestral condition from which those species lacking the exopod developed. Therefore, the species with a one-egg ephippium (all of which lack the exopod on the first leg of the male) represent the more advanced type.

A proposed phylogenetic association of the species is given in fig. 48 using the above mentioned features for relating species.

There is a striking parallel between this phylogeny and the geographical distribution of the species of each line. Species of the *"tenuicornis"* line are almost completely restricted to the Old World, while most of the species belonging to the *"australiensis"* line are found in the New World (fig. 48). There are exceptions to this as will be described below.

Because *Moina tenuicornis* and *australiensis* are the most generalized species of the family it seems probable that they represent the oldest species or at least have retained many primitive characters. These two species now have their centers of distribution in Australia—*tenuicornis* also occurs in South Africa while *australiensis* is found in New Zealand (fig. 20). These two forms either originated in Australia or compose relicts of populations which were once more widely distributed. The latter is suggested by the occurrence of *tenuicornis* in Africa.

The major phylogenetic line represented by *tenuicornis* seems to have spread outward from Africa to all parts of the Old World. *Moina belli*, a species morphologically intermediate to *macrocopa* and *tenuicornis*, is now largely restricted to Africa and part of the Middle East (fig. 31) while *Moina macrocopa* is restricted to the Northern Hemisphere (fig. 4).

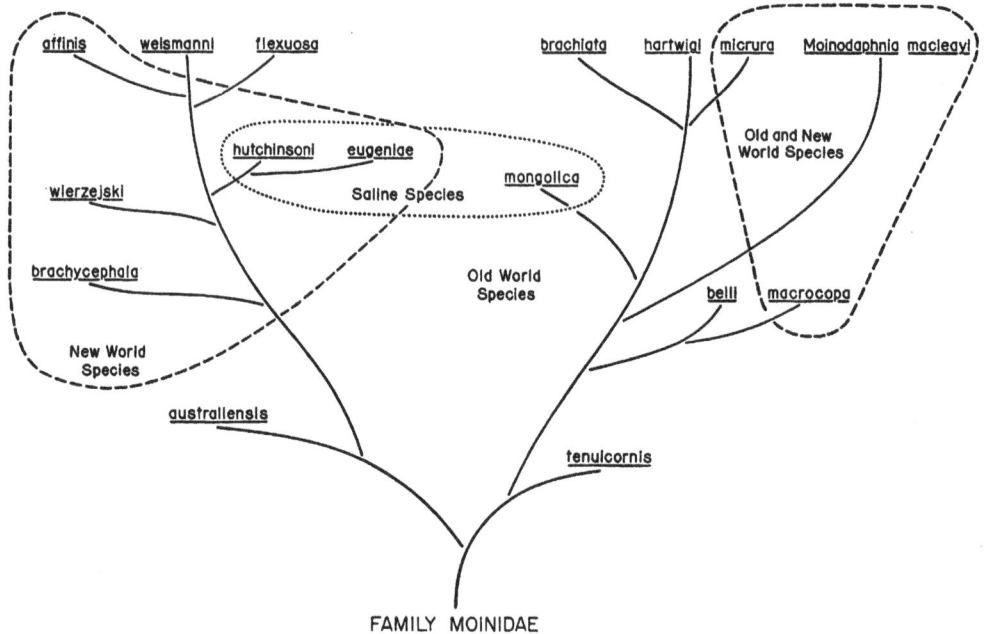

FIG. 48. Phylogenetic interrelationships between species of the Family Moinidae (exclusive of *Moina minuta* and *reticulata*).

These three species are all large forms which produce two sexual eggs. There are also several species of the *"tenuicornis"* group which are small and have only a single sexual egg. Of this latter group, *Moina hartwigi* is restricted to eastern Africa (fig. 31), *Moina brachiata* and *micrura* are found in Africa as well as throughout the Palearctic region (figs. 4 and 13, respectively), and *mongolica* is restricted to very saline pools in Northern Africa, the Middle East, Southern U.S.S.R., and the Far East (fig. 31). *Moinodaphnia macleayi* is found in the humid tropics of Africa and Asia and the coastal areas of Australia (fig. 42).

Of all these species, only *macrocopa*, *micrura*, and *Moinodaphnia* are also found in the New World. *Moina macrocopa* is restricted to the United States; *Moina micrura* and *Moinodaphnia macleayi* occur throughout the warm temperate and tropical regions of the New World. The presence of a fossil ephippium very similar to *Moina micrura* from Miocene deposits of California suggests that the invasion of the New World by species of the *"tenuicornis"* group occurred well before Miocene time.

On the other hand, species of the *"australiensis"* group apparently spread rather early in their development to the New World. The species most like *australiensis* is *brachycephala* which is now restricted

to California (Mojave Desert; fig. 42). *Moina wierzejskii* and *hutchinsoni*—both large species as are *australiensis* and *brachycephala*—are restricted to the New World (figs. 20 and 42). *Moina wierzejskii* is found in the Central and Western United States, in the Caribbean Islands, and in South America. *Moina hutchinsoni* inhabits saline and alkaline pools in Western North America.

Moina affinis is restricted to the United States although it has recently been found in one locality in Italy (fig. 20). On the other hand, *Moina weismanni* and *flexuosa*, two species rather similar to *affinis*, are restricted to the Old World. *Moina weismanni* is found in the Far East while *flexuosa* is restricted to Western Australia (fig. 20).

Moina eugeniae, a small species that inhabits saline pools, is found only in Argentina (fig. 42).

The fact that the large, two-sexual egg species of this latter group are, except for *australiensis*, endemic to the New World and not found in the Old World suggests that this group moved at an early date to the New World.

One may deduce patterns of movement for individual species of each group but until a more complete fossil record is available it would be difficult to know what present populations represent only relicts of a formerly widely distributed species. The information available is sufficient, however, to demonstrate how widely separated most species of the two phylogenetic lines are at the present time.

Unfortunately we do not have sufficient information about *Moina minuta* and *reticulata* to indicate to what phylogenetic lines they belong. The two species are found only in South and Central America—*minuta* inhabits estuarine environments along the eastern coast of Central and South America while *reticulata* has been collected only from inundation ponds in Paraguay. Their restriction to the New World may indicate that they belong to the *"australiensis"* group of species but the general form of the parthenogenetic females suggests that this may not be so.

The *"tenuicornis"* line probably represents the older of the two because the species of this group are most divergent from one another while the species of the *"australiensis"* line still share many characters.

Geographical isolation has undoubtedly been a very effective factor in speciation within the Moinidae. The origin of all species of the group can probably be explained by this phenomenon or by the specific adaptation to new types of habitats particularly permanent freshwater or saline lakes.

BIBLIOGRAPHY

ALLEN, E., and A. M. BANTA. 1929. "Growth and Maturation in the Parthenogenetic and Sexual Eggs of *Moina macrocopa*." *Jour. Morph. and Physiol.* 48: pp. 123–151.

ARÉVALO, C. 1920. "Notas hidrobiológicas." *Bol. Real soc. españ. hist. natur. Madrid* 20: pp. 163–167.

ARORA, G. L. 1931. "Fauna of Lahore." *Bull. Dept. Zool., Panjab Univ.* 1: pp. 62–100.

BAIRD, W. 1850. *The Natural History of the British Entomostraca* (London, Ray Society).

BANTA, A. M. 1939. "Studies on the Physiology, Genetics, and Evolution of some Cladocera." *Carnegie Institution of Washington, Department of Genetics*, Paper No. 39.

BÄR, G. 1924. "Uber Cladoceren von der Insel Ceylon." *Jena. Zeitschr. Naturwiss.* 60: pp. 83–126.

BEHNING, A. L. 1941. *Cladocera of the Caucasus* (Tbilis). (In Russian.)

BERG, K. 1929. "A Faunistic and Biological Study of Danish Cladocera." *Viden. Medd. fra Dansk natur. Forening* 88: pp. 31–111.

—— 1931. "Studies on the Genus *Daphnia* O. F. Müller with Especial Reference to the Mode of Reproduction." *Viden. Medd. fra Dansk natur. Forening* 92: pp. 1–222.

BINDER, G. 1932. "Das Muskelsystem von *Daphnia*." *Internat. Rev. Hydrobiol.* 26: pp. 54–111.

BIRGE, E. A. 1893. "Notes on Cladocera. III.'" *Trans. Wisconsin Acad. Sci.* 9: pp. 275–317.

—— 1918. "The Water Fleas (Cladocera)." H. B. Ward and G. C. Whipple, eds., *Freshwater Biology* (New York, John Wiley and Sons), chap. 22: pp. 676–740.

BIRABÉN, M. 1917. "Nota sobre des Cladóceros nuevos de la República Argentina." *Physis* 3: pp. 262–266.

—— 1919. "Sobre algunos Cladóceros de la Republica Argentina." *Rev. Mus. La Plata* 24: pp. 82–126.

BLANCHARD, R., and J. RICHARD. 1892. "Faune des lacs sales d'Algerie, Cladoceres et Copepodes." *Mem. Soc. Zool. de France* 4: pp. 512–535.

BRADY, G. S. 1886. "Notes on Entomostraca Collected by Mr. A. Haly in Celon." *Jour. Linn. Soc., London, Zool.* 19: pp. 293–317.

BREHM, V. 1933. "Die Cladoceren der Deutschen Limnologischen Sunda-Expedition." *Arch. Hydrobiol.*, Suppl. 11: pp. 631–771.

—— 1935. "Crustacea. I. Cladocera und Euphyllopoda." *Mission Scientifique de L'Omo* 2: pp. 141–166.

—— 1936. "Report on Cladocera." *Mem. Connecticut Acad.* 10: pp. 283–297.

—— 1937a. "Zwei neue Moina-Formen aus Nevada, USA." *Zool. Anz.* 117: pp. 91–96.

—— 1937b. "Cladoceren aus Palästina." *Zool. Anz.* 120: pp. 21–23.

—— 1938. "Über die Süsswasserfauna von Kurdistan." *Zool. Anz.* 121: pp. 209–219.

—— 1948. "Datos para la fauna de agua dulce de Cuba." *P. Inst. Biol. Apl.* 5: pp. 95–112.

—— 1951. "Cladocera und Copepoda Calanoida von Cambodja." *Cybium* 6: pp. 95–124.

—— 1953a. "Indische Diaptomiden, Pseudodiaptomiden und Cladoceren." *Österreichische Zool. Zeit.* 4: pp. 241–345.

—— 1953b. "Report No. 2 from Professor T. Gislen's Expedition to Australia in 1951–1952. Bericht über Cladoceren und Copepoden." *Acta Univ. Lund.*, N. F., 49, 7: pp. 1–11.

—— 1958. "Die Cladocerenfauna des Tassili Gebietes." *Trav. Inst. Rech. Sahariennes (Ser. Tassili)* 3: pp. 73–84.

—— 1963. "Einige Bemerkungen zu vier indischen Entomostraken." *Internat. Rev. Hydrobiol* 48: pp. 159–172.

BROOKS, J. L. 1957. "The Systematics of North American *Daphnia*." *Mem. Connecticut Acad.* 13: pp. 1–180.

—— 1959. "Cladocera." H. B. Ward and G. C. Whipple, eds., *Freshwater Biology* (2nd ed.; by W. T. Edmondson; New York, John Wiley and Sons), chap. 27: pp. 587–656.

—— 1965. "Predation and Relative Helmet Size in Cyclomorphic *Daphnia*." *Proc. Nat. Acad. Sci.* 53: pp. 119–126.

BROOKS, J. L., and S. I. DODSON. 1965. "Predation, Body Size, and Composition of Plankton." *Science* 150: pp. 28–35.

CANNON, H. G. 1933. "VIII. On the Feeding Mechanism of the Branchiopoda." *Phil. Trans. Roy. Soc., London, B* 222: pp. 267–352.

DADAY, J. 1883. "Adatok a Szent-Anna es Mohosto faunajanak ismeretehez." *Orvos-termes., Ertesito* 5.

—— 1888. *Crustacea Cladocera Faunae Hungaricae* (Budapest).

—— 1901. "Mikroskopische Süsswasserthiere." *Horvath's Zool. Ergebn. Zichy* 2: pp. 375–470.

—— 1905. "Untersuchungen über die Süsswasser-mikrofauna Paraguays." *Zoologica* 44: pp. 1–374.

—— 1910. "Untersuchungen über die Süsswasser-mikrofauna Deutsch-Ost-Afrikas." *Zoologica* 59: pp. 1–314.

—— 1928. "Cladoceren und Ostracoden aus Süd-und Südwestafrika." *Denks. Med. Naturw. Ges. Jena* 5: pp. 89–102.

DEHN, M. 1948. "Experimentelle Untersuchungen über den Generationswechsel der Cladoceren. II. Cytologische Untersuchungen bei *Moina rectirostris*." *Chromosoma* 3: pp. 167–194.

DELACHAUX, T. 1917. "Cladoceres de la region du lac Victoria Nyanza." *Rev. Suisse Zool.* 25: pp. 77–93.

ERIKSSON, S. 1934. "Studien über die Fangapparate der Branchiopoden nebst einigen phylogenetischen Bemerkungen." *Zool. Bidr. Uppsala* 15: pp. 23–287.

EYLMANN, E. 1887. "Beitrag zur Systematik der europäischen Daphniden." *Ber. Naturf. Ges. Freiburg* 21: pp. 1–102.

FADEEV, N. N. 1925. "Information about the Fauna of the Lakes of the Transcaucasus." *Rabot. Severo-Kavkak. Hudrobiol. Stat.* 1: pp. 17–26. (In Russian; not seen, in Behning, 1941.)

FISCHER, S. 1851. "Bemerkungen über einige weniger genau gekannte Daphnienarten." *Bull. Soc. Nat. Moscou* 24: pp. 96–108.

FOX, H. M. 1951. "Proposed Suppression under the Plenary Powers of the Generic Name "*Monoculus*" Linnaeus, 1758. "*Bull. Zool. Nomenclat.* 2: pp. 37–39.

FRYER, G. 1963. "The Functional Morphology and Feeding Mechanism of the Chydorid Cladoceran "*Eurycerous lamellatus*" (O. F. Müller). *Trans. Roy. Soc. Edinburgh* 65: pp. 335–381.

GAUTHIER, H. 1954. *Essai sur la variabilité l'ecologie, le déterminisme du sexe et la reproduction de quelques Moina (Cladocères) récoltées en Afrique et a Madagascar* (Alger).

GROSVENOR, G. H., and G. SMITH. 1913. "The Life Cycle of *Moina rectirostris*." *Quart. Jour. Microscop. Sci.* 58: pp. 511–522.

GRUBER, A., and A. WEISMANN. 1880. "Ueber einige neue oder unvollkommen gekannte Daphniden." *Ber. Freiberg Naturf. Ges.* 7: pp. 50–116.

GUERNE, J. DE, and J. RICHARD. 1892. "Cladocères et Copépodes d'eau douce des environs de Rufisque." *Mém. Soc. Zool. France* 5: pp. 526–538.

GURNEY, R. 1904. "On a Small Collection of Freshwater Entomostraca from South Africa." *Proc. Zool. Soc. London* 2: pp. 298–301.

—— 1909. "On the Freshwater Crustacea of Algeria and Tunisia." *Jour. Royal Microscop. Soc.* 1909: pp. 273–305.

—— 1911. "On some Freshwater Entomostraca from Egypt and the Soudan." *Ann. Mag. Nat. Hist.* 8: pp. 28–33.

—— 1927. "Some Australian Freshwater Entomostraca Reared from Dried Mud." *Proc. Zool. Soc. London* 1927: pp. 59–79.

HANSEN, H. J. 1899. "Die Cladoceren und Cirripedien der Plankton-Expedition." *Ergeb. Atlantic Ocean Plankton-Exped. Humboldt-Stiftung* 2: pp. 1–58.

HANTSCHMANN, S. 1961. "Active Compensation for the Pull of Gravity by a Planktonic Cladoceran, *Daphnia schoedleri* Sars." Ph.D. Thesis, Yale University.

Harding, J. P. 1957. "Crustacea: Cladocera." *Res. Sci. Explor. Hydrobiol. Lac Tanganika* 3: pp. 53–89.

HELLICH, M. B. 1877. *Die Cladoceren Böhmens* (Prague).

HEMMING, F. 1958. "Official Index of Rejected and Invalid Generic Names in Zoology" (London, Internat. Trust Zool. Nomenclat.).

HEMSEN, J. 1952. "Ergebnisse der Österreichischen Iran-Expedition 1949/50. Cladoceren und freilebende Copepoden der Kleingewässer und des Kaspisees." *Sitz. Öst Akad. Wiss.* 161: pp. 585–644.

HENRY, M. 1922. "A Monograph of the Freshwater Entomostraca of New South Wales." *Linn. Soc. New South Wales* 47: pp. 26–52.

HERRICK, C. L. 1887. "Contribution to the Fauna of the Gulf of Mexico and the South." *Mem. Denison Sci. Assoc.* 1: pp. 1–56.

HRBÁČEK, J. 1959. "Density of the Fish Population as a Factor Influencing the Distribution and Speciation of the Species in the Genus *Daphnia*." *XVth Internat. Cong. Zool.*, *Sect. X, Paper* 27: pp. 1–2.

HUDENDORFF, A. 1876. "Beitrag zur Kenntniss der Süsswasser Cladoceren Russlands." *Bull. Soc. Nat. Moscou* 50: pp. 26–61.

HUTCHINSON, G. E. 1940. "A Revision of the Corixidae of India and Adjacent Regions." *Trans. Connecticut Acad.* 33: pp. 339–476.

International Code of Zoological Nomenclature Adopted by the XVth International Congress of Zoology. (London, 1961).

ISHIKAWA, C. 1896. "Phyllopod Crustacea of Japan." *Zool. Mag. Tokyo* 8: pp. 1–6.

JENKIN, P. M. 1934. "Cladocera from the Rift Valley Lakes in Kenya." *Ann. Mag. Nat. Hist.*, Ser. 10, 13: pp. 137–160, 281–308.

JOBLOT, 1754. *Observations d'histoire naturelle faites avec le microscope* (Paris).

JUDAY, C. 1916. "Limnological Studies on some Lakes in Central America." *Trans. Wisconsin Acad. Sci.* 18: pp. 214–250.

JURINE, L. 1820. *Histoire des Monocles qui se trouvent aux environs de Genéve* (Geneve).

KEILHACK, L. 1909. "Phyllopoda." In Brauer (ed.) *Süsswasserfauna Deutschlands*, 10: pp. 1–112.

—— 1914. "Faunistische, systematische und nomenklatorische Bemerkungen über Dauphiné-Entomostraken." *Arch. Hydrobiol.* 9: pp. 150–156.

KEISER, N. 1931. "Über die Cladoceren und Copepoden der Wüste Kara-Kum." *Internat. Rev. Hydrobiol.* 25: pp. 355–372.

KING, R. L. 1853. "On Some of the Species of Daphniadae Found in New South Wales." *Roy. Soc. van Diemens-Land, Papers* 2: pp. 243–263.

KURZ, W. 1874. "Dodekas neuer Cladoceren nebst einer kurzen Üebersicht der Cladocerenfauna Böhmens." *Sitz. Acad. Wiss. Wien, Math. Naturw.* 70: pp. 7–88.

LEYDIG, F. 1860. *Naturgeschichte der Daphniden (Crustacea, Cladocera).* (Tübingen).

LIEVEN. 1848. "Die Branchiopoden der Danziger Gegend." *Neuest. Schrift. Naturforsch. Gesellsch. Danzig* 4: pp. 1–52.

LILLJEBORG, W. 1853. *De Crustaceis ex ordinibus tribus: Cladocera, Ostracoda et Copepoda, in Scania Occurrentibus* (Lund).

—— 1900. "Cladocera Sueciae." *Königl. Ges. Wiss. Upsala.* 6: pp. 1–701.

LÖFFLER, H. 1956. "Ergebnisse der Österreichischen Iranexpedition 1940/50: Limnologische Beobachtungen an Iranischen Binnengewässern." *Hydrobiol.* 8: pp. 201–278.

—— 1961. "Beiträge zur Kenntnis der Iranischen Binnengewässer. II. Regional-limnologische Studie mit besonderer Berücksichtigung der Crustaceenfauna." *Internat. Rev. Hydrobiol.* 46: pp. 309–406.

MARCHI, M. 1913. "Notizia sulla presenza di *Moina rectirostris* (F. Leydig) nel Trentino." *Rendic. R. Ist. Lombardo. Sci. Lett.* 46: pp. 811–821.

MATILE, P. 1891. "Die Cladoceren der Umgegend von Moskau." *Bull. Soc. Natur. Moscou*, Nouv. Ser. 4: pp. 104–169.

MONIEZ, R. 1888. *Matériaux pour sevir á l'étude de la faune des eaux douces des Acores. IV. Crustacés: Cladocéres* (Lille). (Not seen.)

MORONI, A. 1962. *L'Ecosistema di Risaia.* Milano Ente Naz. Risi.

MORONI, A., and E. VICINI. 1962. "Ulteriori ricerche sull' ecosistema di risaia." *L'Ateneo Parmense* 33: pp. 1–35.

MÜLLER, P. E. 1868. "Danmarks Cladocera." *Naturh. Tidssk.* 5: pp. 53–240.

MÜLLER, O. F. 1776. *Zoologiae Danicae Prodromus* (Havniae).

—— 1785. *Entomostraca seu insecta testacea, quae in aquis Daniae et Norvegiae reperit, descripsit et iconibus illustravit.* (Havniae).

MURAKAMI, Y. 1961. "Studies on the Winter Eggs of the Water Flea *Moina macrocopa* Straus." *Jour. Facult. Fish., Hiroshima Univ.* 3: pp. 323–346.

OLIVIER, S. R. 1954. "Una nueva especie del género ⟨⟨*Moina*⟩⟩ (Crust., Cladócera)." *Not. Mus. Eva Peron* 17: pp. 81–86.

—— 1962. "Los Cladoceres Argentinos con claves de las especies, notas biologicas y distribucion geografica." *Rev. Mus. La Plata*, N. S. Zool., 7: pp. 173–269.

PACAUD, A. 1952. "Remarques sur la Systématique du Genre *Moina* Baird (Cladocères) et sur sa distribution autour du Bassin Occidental de la Méditerrainée." *Vie et Milieu* 3: pp. 68–76.

PALMER, A. R. 1957. "Miocene Arthropods from the Mojave Desert, California." *Geol. Surv., Prof. Paper* 294–G: pp. 237–280.

—— 1960. "Miocene Copepods from the Mojave Desert, California." *Jour. Paleont.* **34**: pp. 447–452.

PIROCCHI, L. 1940. "Planctonti di pozze d'alpeggio." *Mem. Mus. Stor. Natur. Venezia Trid.* **5**: pp. 1–25.

RAMMNER, W. 1931. "Ein Vorkommen von *Moina dubia* Guerne and Richard in Deutschland." *Arch. Hydrobiol.* **22**: pp. 623–634.

—— 1933. "Zoologische Ergebnisse einer Reise nach Bonaire, Curacao und Aruba im Jahre 1930. No. 4. Süss—u. Brackwasser—Phyllopoden von Bonaire." *Zool. Jahrb.* **64**: pp. 357–368.

—— 1937. "Beitrag zur Cladocerenfauna von Java." *Internat. Rev. Hydrobiol.* **35**: pp. 35–50.

RICHARD, J. 1888. "Note sur *Moina bathycola* (Vernet)." *Zool. Anz.* **11**: pp. 118–119.

—— 1891. "Entomostracés d'eau douce de Sumatra et de Celebes." *Zool. Ergbn. Reise Niederl. Ost-Indien* **2**: pp. 118–128.

—— 1892. "*Grimaldina Brazzai, Guernella Raphaelis, Moinodaphnia Mocquerysi,* Cladocéres nouveaux du Congo." *Mem. Soc. Zool. France* **5**: pp. 213–226.

—— 1895. "Sur quelques Entomostracés d'eau douce d'Haiti." *Mem. Soc. Zool. France* **8**: pp. 189–199.

ROBIN, C. 1872. "Observations anatomiques et zoologiques sur deux especes de Daphnies." *Jour. L'Anat. et Physiol.* **8**: pp. 449–467.

RÜHE, F. E. 1914. "Die Süsswassercrustaceen der Deutschen Südpolar-Expedition 1901–1903." *Deutsche Südpolar Exped., Zool.* **8**: pp. 5–66.

SARS, G. O. 1865. "Norges Ferskvandskrebsdyr Förste Afsnit Branchiopoda. I. Cladocera Ctenopoda." *Norges Ferskvandskreb.* 1865: pp. 1–71.

—— 1885. "On some Australian Cladocera Raised from Dried Mud." *Forh. Vidensk.-Selsk., Cristiania* 1885: pp. 1–46.

—— 1890. "Oversigt af Norges Crustaceer. II. (Branchiopoda, Ostracoda, Cirripedia)." *Forh. Vidensk.-Selsk., Cristiana* 1890: pp. 1–80.

—— 1896. "On Fresh-water Entomostraca from the Neighbourhood of Sydney, Partly Raised from Dried Mud. *Arch. Math. Naturviden.* **18**: pp. 1–81.

—— 1897. "On some West-Australian Entomostraca." *Arch. Math. Naturviden.* **19**: pp. 1–35.

—— 1901. "Contributions to the Knowledge of the Fresh-water Entomostraca of South America, as shown by Artificial Hatching from Dried Material. *Arch. Math. Naturviden.* **23**: pp. 1–102.

—— 1903a. "On the Crustacean Fauna of Central Asia. Part II. Cladocera." *Ann. Mus. St. Petersburg.* **8**: pp. 157–194.

—— 1903b. "Fresh-water Entomostraca from China and Sumatra." *Arch. Math. Naturviden.* **25**: 1–44.

—— 1916. "The Fresh-water Entomostraca of Cape Province (Union of South Africa). Part I. Cladocera." *Ann. South African Mus.* **15**: pp. 303–351.

SCHIKLEJEW, S. M. 1930. "Die Cladocerafauna der Kaukasischen Hegegewässer und des Strandes des Schwarzen Meeres." *Arch. Hydrobiol.* **21**: pp. 336–349.

SCHOEDLER, J. E. 1877. "Zur Naturgeschichte der Daphniden." *Abh. Dorotheenstädt. Realsch.* **77**: pp. 1–24.

SCOURFIELD, D. J. 1903. "A Synopsis of the Known Species of British Freshwater Entomostraca. I. Cladocera." *Jour. Quekett Microscop. Club* **8**: pp. 431–454.

SCOURFIELD, D. J., and J. P. HARDING. 1941. "A Key to the British Species of Freshwater Cladocera with Notes on Their Ecology." *Sci. Pub. Freshw. Biol. Assoc., Ambleside* **5**: pp. 1–50.

SCOURFIELD, D. J., and J. P. HARDING. 1958. "A Key to the British Species of Freshwater Cladocera." *Sci. Pub. Freshw. Biol. Assoc., Ambleside* (2nd ed.) **5**: pp. 1–55.

SPANDL, H. 1926. "Das Zooplankton des Paranagua-Sees (Brazil)." *Denk. Akad. Wiss. Wien* **76**: pp. 101–105.

ŠRÁMEK-HUŠEK, R. 1940. "K systematice a oekologii perloocky *Moina micrura* Kurz a ostatnich Moin v. Cechach." *Casopis Narodniho Musea V, Praze* **114**: pp. 204–214.

ŠRÁMEK-HUŠEK, R., M. STRAŠKRABA, J. BRTEK. 1962. "Branchiopoda." *Fauna ČSSR*" **16**: pp. 1–470.

STEPANOV, P. 1885. "Fauna vom Wesowsee." *Arb. Charkow Natf. Ges.* **19**: pp. 13–43.

—— 1886. "Material for the Study of the Fauna of the Saline Lakes of Slavyansk." *Bull. Soc. Nat. Moscou* **62**: pp. 185–199. (In Russian.)

STEPHANIDES, T. 1936. "On the Presence in Corfu of *Moina salinarum* Gurney and *Moina belli* Gurney." *Arch. Hydrobiol.* **29**: pp. 691–694.

—— 1937. "On the Ephippium of *Moina belli* Gurney." *Arch. Hydrobiol.* **31**: pp. 163–164.

—— 1948. "A Survey of the Fresh-water Biology of Corfu and of Certain Other Regions of Greece." *Prakt. Hellinic Hydrobiol. Inst.* **2**: pp. 1–263.

STEUER, A. 1939. "Variabilität und Verbreitung von *Moina dubia* Guerne u. Richard." *Sitz. Akad. Wiss. Wien* **148**: pp. 269–278.

STINGELIN, T. 1900. "Beitrag zur Kenntnis der Süsswasserfauna von Celebes." *Rev. Suisse Zool.* **8**: pp. 193–207.

—— 1904. "Entomostraken, gesammelt von Dr. G. Hagmann in Mündungsgebiet des Amazonas." *Zool. Jahr.* **20**: pp. 575–590.

—— 1914. "Cladoceren aus den Gebirgen von Kolumbien." *Mem. Soc. Neuchatel. Sci. Nat.* **5**: pp. 600–638.

STORCH, O. 1926. "Morphologie und Physiologie des Fangapparates der Daphniden." *Ergebn. Fort. Zool.* **6**: pp. 125–234.

STRAUS, H. E. 1819. "Mémoire sur les Daphnia, de la classe des Crustacés." *Mem. Mus. D'Hist. Natur.* **5**: pp. 380–425.

—— 1820. "Mémoire sur les Daphnia, de la classe des Crustacés (Secondi Partie)." *Mem. Mus. D'Hist. Natur.* **6**: pp. 149–162.

STUHLMANN, F. 1889. "Zweiter Bericht über eine mit Unterstützung der Königlichen Akademie der Wissenschaften nach Ost-Africa unternommene Reise." *Sitz. Königlich Preuss. Akad. Wiss. Berlin* 1889: pp. 645–660.

TSI-CHUNG, C., and L. S. CLEMENTE. 1954. "The Classification and Distribution of Fresh Water Cladocerans Around Manila." *Nat. Appl. Sci. Bull.* **14**: pp. 85–150.

UENO, M. 1927. "The Freshwater Branchiopoda of Japan." *Mem. Coll. Sci., Kyoto Imp. Univ.*, B **2**: pp. 259–311.

—— 1936. "Cladocera of Lake Ngardok in Babelthaop of the Palau Islands." *Annot. Zool. Japonenses* **15**: pp. 514–519.

—— 1937. "Cladocera of Manchoukuo." *Internat. Rev. Hydrobiol.* **35**: pp. 199–216.

—— 1939. "Manchurian Freshwater Cladocera." *Annot. Zool. Japonenses* **18**: pp. 219–231.

VERESTCHAGIN, G. 1914. "Some Remarks on the Fauna of Entomostraca of Central Africa." In Dogiel and Sokolov (eds.). *Sci. Res. Zool. Exp. Brit. E. Africa and Uganda* **1**: pp. 1–27. (In Russian.)

VERNET, H. 1878. "Entomostracées de la faune profonde du Lac Leman." *Bull. Soc. Vaud.* **15**: pp. 526–535. (Not seen; in Richard, 1888.)

WAGLER, E. 1937. "Crustacea." *Die Tierwelt Mitteleuropas* **2**: pp. 1–224.

WEISMANN, A. 1877a. "Beiträge zur Naturgeschichte der Daphnoiden. III. Die Abhängigkeit der Embryonal-Entwicklung vom Fruchtwasser der Mutter." *Zeit. Wissen. Zool.* **28**: pp. 176–211.

—— 1877b. "Beiträge zur Naturgeschichte der Daphnoidan. IV. Ueber den Einfluss der Begattung auf die Erzeugung von Wintereiern." *Zeit. Wissen. Zool.* **28**: pp. 212–240.

—— 1879a. "Beiträge zur Naturgeschichte der Daphnoidan. VI. Samen und Begattung der Daphnoiden." *Zeit. Wissen. Zool.* **33**: pp. 55–110.

—— 1879b. "Beiträge zur Naturgeschichte der Daphnoidan. VII. Die Entstehung der cyclischen Fortpflanzung bei den Daphnoiden." *Zeit. Wissen. Zool.* **33**: pp. 111–270.

WEISMANN, A., and C. ISCHIKAWA. 1891. "Ueber die Paracopulation in Daphnidenei, sowei über Reifung und Befruchtung desselben." *Zool. Jahrb.* **4**: pp. 155–196.

WELTNER, W. 1898. "Ostafrikanische Cladoceren ges. v. Dr. Stuhlmann 1888 u. 1889." *Mitteil. Naturh. Mus. Hamburg* **15**: pp. 133–144.

WIERZEJSKI, A. 1892. "Skorupiaki i wrotki (Rotatoria) slodkowodne zebrane w Argentynie." *Rozp. Wyd. mat. przyr. Akad. Umiéj. Krakowie*, ser. II, **4**: pp. 229–246.

WORTHINGTON, E. B., and C. K. RICARDO. 1936. "Scientific Results of the Cambridge Expedition to the East African Lakes 1930. No. 17. The vertical distribution and movements of the plankton in Lakes Rudolf, Naivasha, Edward, and Bunyoni." *Jour. Linn. Soc. London, Zool.* **40**: pp. 33–69.

ZENKEVITCH, L. 1963. *Biology of the Seas of the U.S.S.R.* (New York, John Wiley and Sons, Inc.).

INDEX

99

www.ingramcontent.com/pod-product-compliance
Lightning Source LLC
Chambersburg PA
CBHW081337190326

41458CB00018B/6026